Springer Theses

Recognizing Outstanding Ph.D. Research

Aims and Scope

The series "Springer Theses" brings together a selection of the very best Ph.D. theses from around the world and across the physical sciences. Nominated and endorsed by two recognized specialists, each published volume has been selected for its scientific excellence and the high impact of its contents for the pertinent field of research. For greater accessibility to non-specialists, the published versions include an extended introduction, as well as a foreword by the student's supervisor explaining the special relevance of the work for the field. As a whole, the series will provide a valuable resource both for newcomers to the research fields described, and for other scientists seeking detailed background information on special questions. Finally, it provides an accredited documentation of the valuable contributions made by today's younger generation of scientists.

Theses are accepted into the series by invited nomination only and must fulfill all of the following criteria

- They must be written in good English.
- The topic should fall within the confines of Chemistry, Physics, Earth Sciences, Engineering and related interdisciplinary fields such as Materials, Nanoscience, Chemical Engineering, Complex Systems and Biophysics.
- The work reported in the thesis must represent a significant scientific advance.
- If the thesis includes previously published material, permission to reproduce this must be gained from the respective copyright holder.
- They must have been examined and passed during the 12 months prior to nomination.
- Each thesis should include a foreword by the supervisor outlining the significance of its content.
- The theses should have a clearly defined structure including an introduction accessible to scientists not expert in that particular field.

More information about this series at http://www.springer.com/series/8790

Thomas Keck

Machine Learning at the Belle II Experiment

The Full Event Interpretation and Its Validation on Belle Data

Doctoral thesis accepted by the Karlsruhe Institute of Technology, Karlsruhe, Germany

Springer

Author
Dr. Thomas Keck
Institute of Experimental Particle Physics
Karlsruhe Institute of Technology
Karlsruhe, Germany

Supervisor
Prof. Michael Feindt
Karlsruhe Institute of Technology
Karlsruhe, Germany

ISSN 2190-5053 ISSN 2190-5061 (electronic)
Springer Theses
ISBN 978-3-319-98248-9 ISBN 978-3-319-98249-6 (eBook)
https://doi.org/10.1007/978-3-319-98249-6

Library of Congress Control Number: 2018950801

This Springer imprint is published by the registered company Springer Nature Switzerland AG
The registered company address is: Gewerbestrasse 11, 6330 Cham, Switzerland

Supervisor's Foreword

In this thesis, Thomas Keck has automated and optimized a large part of the reconstruction work of highly skilled elementary particle physicists at the Belle and Belle II experiments in Japan performed by large international collaborations (450 and >700 researchers from the whole world) by means of a hybrid (expert knowledge and machine learning) hierarchical artificial intelligence system. Based on a first version for the Belle experiment developed in 2011, he improved the performance by another factor of 2 to an overall factor of 4 compared to human performance in reconstruction efficiency at constant purity. He invented several new machine learning methods making possible trainings independent of systematic uncertainties in Monte Carlo simulations. He made it work both for the existing Belle data and the future Belle II data. In addition, he performed an analysis of an important rare decay of B mesons with the potential to find new physics beyond the current standard model. Here, he pointed out previously unknown systematic uncertainties.

The thesis poses an important step in improving the human expert efficiency in exploiting expensive fundamental science experiments by means of artificial intelligence/machine learning (AI/ML). It is a highly enjoyable, readable, successful and valuable application of ML methods and shows real value and progress in elementary particle physics beyond the current hype around AI, a superb example for the new development of data science for science. It also shows that deep human understanding and intelligence is the basis for successful application of artificial intelligence. Notwithstanding my high-level view in this foreword: the thesis is a scientific paper containing excellent reviews and new developments and results about both experimental B flavour physics and machine learning.

Karlsruhe, Germany
July 2018

Prof. Feindt Michael

Acknowledgements

I thank all the people who contributed to the success of this work.

> If I have seen further it is by standing on the shoulders of giants.
>
> —Isaac Newton

I can look back on 4 wonderful years at the Institute of Experimental Particle Physics. Therefore, I would like to thank my colleagues and friends for the wonderful time and the creative environment. Because of you, I looked forward to every single day at the institute and learned something new on each of these days.

I would like to mention the technical contributions of my colleagues separately here and would like to express my special thanks to them in doing so.

I got the idea for a Belle to Belle II data conversion during a discussion with Dr. Bastian Kronenbitter. The starting point for the `belle_legacy` library was a cleaned up version of the `Belle Analysis Software Framework`, which I received from Dr. Ryosuke Itoh.

Dr. Martin Ritter supported me in integrating this library into BASF2 and connecting it to the `Belle II Condition Database`. The `b2bii` package was developed in cooperation with the Belle II `b2bii` task force under the direction of Dr. Anze Zupanc. Among the first users who were particularly important for feedback were Felix Metzner, Moritz Gelb and Markus Prim.

My knowledge of multivariate methods was strongly influenced by Prof. Dr. Michael Feindt. The original ideas for the data-driven methods and for the further development of sPlot technique originate from him. The first implementation of sPlot technology in the Belle II software framework was developed during Benjamin Lipp's bachelor's thesis. I first heard the term deep learning from Dr. Martin Heck. Essential contributions to the integration of deep learning technologies into the MVA package were made during Jochen Gemmler's master's thesis and doctoral thesis, and Dennis Weyland's master's thesis. Nils Braun and Oliver Frost were a driving force in integrating the MVA package into the reconstruction software.

The first version of the FEI was strongly influenced by the full reconstruction, which was developed by many doctoral students at the EKP between 2008 and 2012. This first version was developed in close cooperation with Dr. Christian Pulvermacher. I received many suggestions and the idea for the specific FEI from Dr. Martin Heck. The calibration of the FEI on Belle data was carried out by Judith Schwab in her master's thesis. The first application of the FEI on Belle data was carried out by Felix Metzner in his master's thesis.

The analysis $B^+ \rightarrow \tau^+ \nu_\tau$ is based on the doctoral theses of Dr. Bastian Kronenbitter, Dr. Johannes Grygier and Dr. Michael Ziegler.

The content and language of this work were corrected: Nils Braun, Moritz Gelb, Jochen Gemmler, Dr. Pablo Goldenzweig, Dr. Thomas Hauth, Dr. Martin Heck, Felix Metzner, Dr. William Sutcliffe and Judith Schwab.

Equivalent to these professional contributions, the success of this work is based on the support of my supervisors.

I thank Dr. Martin Heck, who has taken over as deputy head of the Feindt working group and has continued the group with great commitment. I thank Prof. Dr. Güter Quast for taking over the correferat and especially for the motivating discussions. I thank my supervisor and lecturer Prof. Dr. Michael Feindt. During my time as a doctoral student, he has enabled me to gain many valuable experiences that I could not have gained in any other working group.

Contents

Chapter 1
Introduction

Physics deals with the development of quantitative models that describe nature. The most fundamental and precise models at present are: general relativity, which describes gravity as an interaction between the curvature of space-time and matter; and the Standard Model of particle physics (SM), which describes matter and the remaining electromagnetic, weak and strong interaction in the form of relativistic quantum fields.

The current high-energy physics (HEP) research focuses on the precise determination of the 19 free parameters of the SM and the search for new physics phenomena beyond the SM.

The Belle II experiment is part of this effort. It is located at the SuperKEKB electron-positron collider in Tsukuba, Japan. It is designed to perform a wide range of high-precision measurements in all fields of heavy flavour physics, including: B meson decays; B_s^0 meson decays; charm physics; τ lepton physics; hadron spectroscopy; and pure electroweak measurements. These measurements will constrain the parameter space of the SM as well as some of its extensions.

This work focuses on the development of software, in particular machine learning algorithms, to advance scientific progress, and to enable and improve a wide range of physics measurements at Belle II. This thesis summarizes my contributions to the Belle experiment and its successor the Belle II experiment.

Four major topics are covered. The conversion of the data recorded by the Belle experiment into the new data-format used by Belle II in Chap. 2. The integration of state-of-the-art machine learning algorithms and novel data analysis techniques into the Belle II Software Framework (BASF2) in Chap. 3. The development of the `Full Event Interpretation` exclusive tagging algorithm, which is unique to the Belle II experiment in Chap. 4. And the validation of the entire analysis software stack using the benchmark measurement of the branching fraction of the rare decay $\mathrm{B} \rightarrow \tau\nu_\tau$ in Chap. 5.

T. Keck, *Machine Learning at the Belle II Experiment*, Springer Theses,
https://doi.org/10.1007/978-3-319-98249-6_1

Chapter 2
From Belle to Belle II

During this thesis, the full $\Upsilon(4S)$ dataset of the Belle experiment and large amounts of the available Monte Carlo data were converted into the new data-format used by Belle II. Therefore, it is possible to evaluate the reliability and performance of the newly developed analysis methods, in particular data-driven techniques, before a comparable dataset from the Belle II experiment is available.

In the following I give a brief overview of the Belle experiment (Sect. 2.1), its successor the Belle II experiment (Sect. 2.2) and describe the technical aspects of converting the Belle dataset into the Belle II data-format (Sect. 2.3).

2.1 The Belle Experiment

From June 1st 1999 until June 30th 2010, the Belle experiment recorded 988 fb^{-1} of data at the KEKB asymmetric e^+e^- collider [1].

This summary of the KEKB accelerator and the Belle detector is based on [1, 2]. Sections 2.1.1 and 2.1.2 are adapted from my master's thesis [3].

2.1.1 KEKB Accelerator

The KEKB collider was dedicated to B physics and operated in the energy range of the Υ resonances. Its asymmetric beam energies induced a Lorentz boost $\beta\gamma = 0.42$ of the center of mass frame relative to the laboratory system, enabling the precise observation of the time evolution of B meson decays. During its runtime between 1998 and 2010 KEKB achieved the highest instantaneous luminosity of $2.1 \times 10^{34}\,\text{cm}^{-2}\,\text{s}^{-1}$ ever achieved by a collider [1, Sect. 1.3]. The machine parameters of KEKB are summarized in Table 2.2.

T. Keck, *Machine Learning at the Belle II Experiment*, Springer Theses,
https://doi.org/10.1007/978-3-319-98249-6_2

2.1.2 Belle Detector

The sole interaction point (IP) was surrounded by the Belle detector to detect and identify particles produced by the collisions. Like the accelerator, the detector was specifically designed for the precise observation of B meson decays. This includes precise measurement of secondary vertices and good particle identification capabilities.

Figure 2.1 shows a schematic side view of the Belle detector.

Going outwards from the interaction point the Belle detector consisted of:

- a double-walled Beryllium beam pipe with a radius of 20 mm cooled by He gas;
- radiation-hard Bismuth Germanate crystals used as an extreme forward calorimeter (EFC) as well as a beam and luminosity monitor;
- four layers of double-sided Silicon strip detectors (SVD) for precise vertex detection;
- a central drift chamber (CDC), which measured momentum and energy loss of charged particles;
- an Aerogel Cherenkov counter (ACC) system for particle identification (PID);
- a time-of-flight (TOF) detector system with plastic scintillation counters;
- a segmented array of CsI (Tl) crystals with silicon photodiode readout for electromagnetic calorimetry (ECL);
- a superconducting solenoid which provided a homogeneous magnetic field of 1.5 T;
- and an iron support structure, which served as the return path of the magnetic flux and was instrumented with glass-electrode resistive plate counters for K_L and muon detection (KLM).

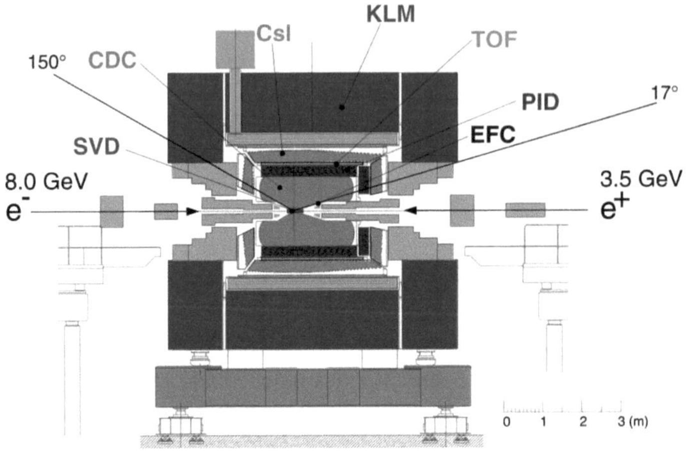

Fig. 2.1 Side view of the Belle detector. Adapted from [2]

Table 2.1 Summary of the integrated luminosity collected by Belle, broken down by center-of-mass energy. Adapted from [1]

Resonance	On-resonance luminosity (fb^{-1})	Off-resonance luminosity (fb^{-1})
$\Upsilon(5S)$	121.4	1.7
$\Upsilon(4S)$ – SVD1	140.0	15.6
$\Upsilon(4S)$ – SVD2	571.0	73.8
$\Upsilon(3S)$	2.9	0.2
$\Upsilon(2S)$	24.9	1.7
$\Upsilon(1S)$	5.7	1.8
Scan > $\Upsilon(4S)$	n/a	27.6

2.1.3 Recorded Dataset

The Belle dataset is grouped into 31 experiments[1]; each experiment marks the time period between two major shutdowns. Experiments 7 to 27 were recorded with the first Silicon Vertex Detector (SVD1). The remainder was recorded with the second (SVD2) detector and, in addition, was reprocessed in 2009 with an improved version of the reconstruction software. Each experiment is further subdivided into a number of runs.

Most of the data was recorded at the center-of-mass energy of the $\Upsilon(4S)$ resonance. In addition, data was also recorded at the $\Upsilon(1S)$, $\Upsilon(2S)$, $\Upsilon(3S)$ and $\Upsilon(5S)$ resonances. Apart from that, off-resonance data 60 MeV below the resonances was collected to estimate the continuum background from data instead of Monte Carlo simulation. Table 2.1 shows a summary of the integrated luminosity collected by Belle, broken down by the center-of-mass energy.

The raw data coming from the detector was calibrated, reconstructed and stored on tape using `PANTHER`-based[2] data summary tape (DST) files. After each experiment the calibration constants were recomputed by detector experts or computed directly from data, and stored in the Belle Condition Database (based on `PostgreSQL`). Finally, the data of the completed experiment was reprocessed and stored in a compact form called mDST files.[3]

For each experiment, ten times the real integrated luminosity in $b\bar{b}$ events and six times that in continuum events were simulated using `EvtGen` and `GEANT3`, and reconstructed with the same software as was used for the detector data.

---

[1]Enumerated from 7 to 73 using only odd numbers and skipping the numbers 29, 57 and 59 for reasons unknown to the author.

[2]The `PANTHER` format consists of tables, which are compressed by the zlib library. The table formats are defined by `ASCII` header files.

[3]A reduced and compressed form of the data summary tape files.

Table 2.2 Achieved parameters of KEKB and design parameters of SuperKEKB [4]

Machine Parameter	KEKB		SuperKEKB	
	e^-	e^+	e^-	e^+
Beam current (A)	1.64	1.19	3.60	2.61
Energy (GeV) (E_{HER}/E_{LER})	8.0	3.5	7.0	4.0
β_y^* (mm)	5.9	5.9	0.27	0.41
Crossing angle (mrad)	22		83	
Beam lifetime (min)	200	150	10	10
Luminosity ($10^{34}\,\text{cm}^{-2}\,\text{s}^{-1}$)	2.11		80	

2.2 The Belle II Experiment

This summary of the SuperKEKB accelerator and the Belle II detector is based on the detailed description in the technical design report [4] and is an updated and condensed version of a summary presented in my master's thesis [3].

2.2.1 *SuperKEKB Accelerator*

The accelerator was shut down in June 2010 and upgraded to SuperKEKB to increase the instantaneous luminosity to $8 \times 10^{35}\,\text{cm}^{-2}\,\text{s}^{-1}$, which is 40 times the peak instantaneous luminosity of KEKB [5, Sect. 2].

The higher instantaneous luminosity is obtained by adopting the Nano-Beam scheme [5, Sect. 2], which requires a larger crossing angle to fit the final focusing magnets [4, Sect. 3.1]. Moreover, the beam current in both rings is doubled. The beam energy asymmetry was reduced to mitigate the shortened beam lifetime due to the Touschek effect.[4] In consequence the Lorentz boost is reduced to $\beta\gamma = 0.28$. The relevant machine parameters of KEKB and SuperKEKB are summarized in Table 2.2.

2.2.2 *Belle II Detector*

The original detector is currently upgraded to match the higher instantaneous luminosity of SuperKEKB. The most important objectives for the upgraded detector are higher physics and background rate tolerance, better physics performance despite smaller Lorentz boost, and improved radiation hardness [5].

[4]The Touschek effect is a loss mechanism due to large angle coulomb scattering inside a bunch.

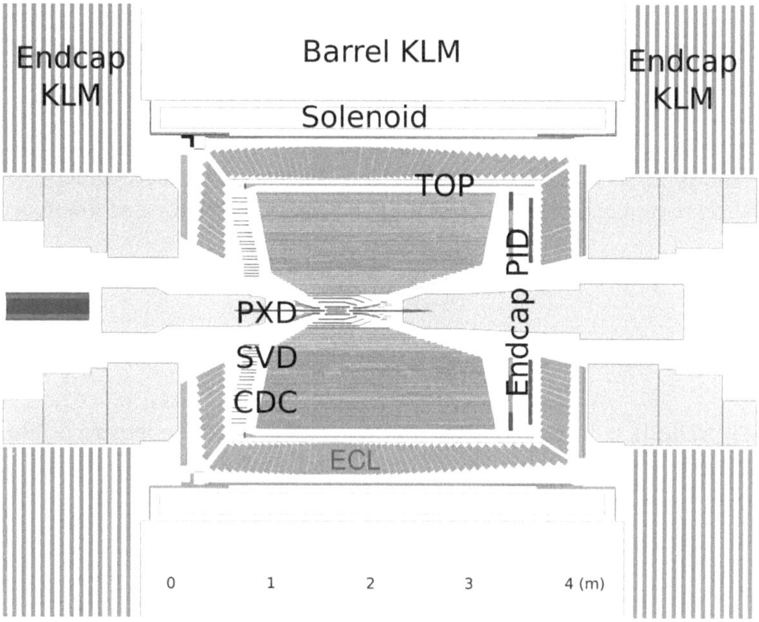

Fig. 2.2 Side view of the upgraded Belle II detector [6]

Figure 2.2 shows the Belle II detector.

Going outwards from the interaction point the Belle II detector consists of:

- a double-walled Beryllium beam pipe with a radius of 12 mm cooled by paraffin;
- a pixel detector based on the DEPFET[5] technology (PXD) for precise vertex detection;
- four layers of double-sided silicon strip detectors covering the full $17° - 150°$ acceptance of the Belle II detector (SVD) to extrapolate the tracks reconstructed in the CDC to the PXD and to reconstruct low-momentum tracks;
- a central drift chamber (CDC), which measures momentum and energy loss of charged particles, and provides a fast trigger signal for the Level-1 (L1) trigger system;
- a proximity-focusing[6] Aerogel Ring-Imaging Cherenkov detector (ARICH) for particle identification (PID) in the forward end-cap;
- a time-of-propagation counter (TOP) in the barrel region using an array of 16 quartz bars between the outer CDC cover and the calorimeter's inner surface providing PID information and a timing signal with a resolution of a few nanoseconds to the trigger system;

[5]DEPleted Field Effect Transistor.

[6]An increasing refractive index is used to reduce the spread of the ring image due to emission point uncertainty.

- a segmented array of CsI(Tl) crystals in the barrel region and pure CsI crystals in the end-caps with photodiode readout for electromagnetic calorimetry (ECL);
- a superconducting solenoid which provides a homogeneous magnetic field of approximately 1.5 T;
- and an iron support structure, which serves as the return path of the magnetic flux, and is instrumented with glass-electrode resistive plate counters (RPC) in the barrel region and scintillator strip in the end-caps for K_L and muon detection (KLM).

2.2.3 Anticipated Dataset

By 2025, Belle II will record 50 ab^{-1} of data, which corresponds to 50 times the integrated luminosity of Belle. The current (March 17th 2017) luminosity projection is shown in Fig. 2.3.

2.2.3.1 Data Acquisition

Belle II uses a two-level trigger system, with an FPGA-based Level-1 (L1) trigger decision and a high level trigger (HLT) farm.

At the design luminosity the nominal average L1 trigger rate is expected to be up to 30 kHz. For each trigger signal the data is read out by the data acquisition system

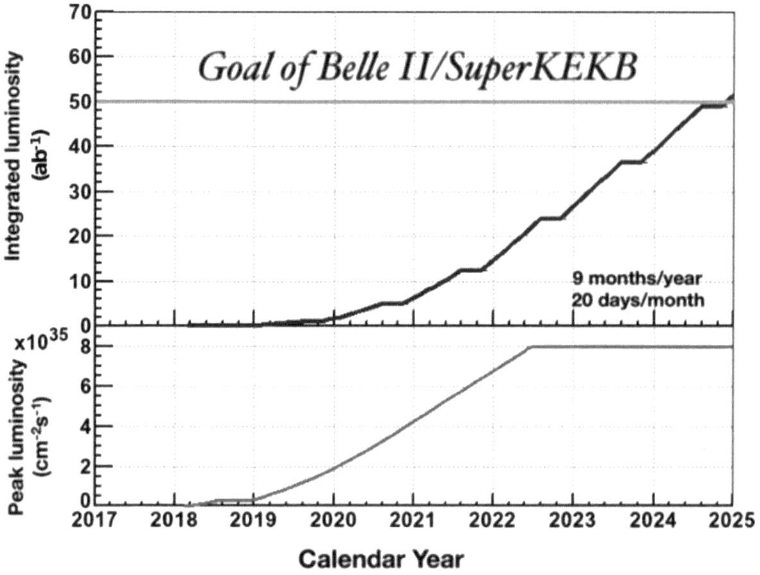

Fig. 2.3 SuperKEKB integrated and instantaneous luminosity projection [7] (March 17th 2017)

from the detector front-end electronics. The high level trigger farm reconstructs the event and performs a physics-level event selection using the event data from all sub-detectors. The final rate of raw events written to the storage is expected to be between 6 and 10 kHz. The HLT consists of multiple Linux-based PC clusters and runs the `Belle II Analysis Software Framework` (`BASF2`) [8].

2.2.3.2 Offline Reconstruction and Monte Carlo Production

The same software framework is used in offline reconstruction, Monte Carlo production, and physics analysis. After machine-dependent calibration parameters are determined, the raw data is reconstructed and stored at the KEK computing center. Monte Carlo production and reconstruction will be distributed to data centers around the world. The reconstructed information is stored in `ROOT`-based mDST files.

2.3 Belle to Belle II Dataset

In the above discussion of the recorded Belle and anticipated Belle II dataset, four levels of data processing can be distinguished:

1. **online reconstruction**—the read-out of the detector and the trigger system, producing the **raw-data** (DST files);
2. **offline reconstruction**—cluster reconstruction, track finding and fitting, producing the **mDST data**;
3. **mDST analysis**—creation of final state particle hypotheses, reconstruction of intermediate particle candidates and vertex fitting, producing **flat n-tuples**;
4. and **n-tuple analysis**—fit to theoretical predictions in order to extract interesting observables, producing **scientific papers**.

Converting the raw-data is in principle possible, but the differences between the Belle and Belle II detector render this a difficult and ill-defined task. While this would allow for the validation of the Belle II reconstruction software (e.g. the track finding and fitting algorithms) on Belle data, this would be only of limited use due to the vastly different expected background and the availability of data from cosmic runs.

The Belle to Belle II dataset conversion converts the Belle mDST data, which contains mostly detector independent objects like tracks and energy clusters, into the new mDST format used by `BASF2`. This enables the validation of the Belle II analysis software, and (re-)production of Belle measurements using the improved software. Figure 2.4 displays an overview of `BASF2` including the conversion path presented in this work.

By comparing the original Belle results, the Belle results obtained from converted data in `BASF2`, and Belle II sensitivity studies on Belle II Monte Carlo, it is possible to assign improvements in the sensitivity and occurring inconsistencies to the analysis and reconstruction algorithms, separately.

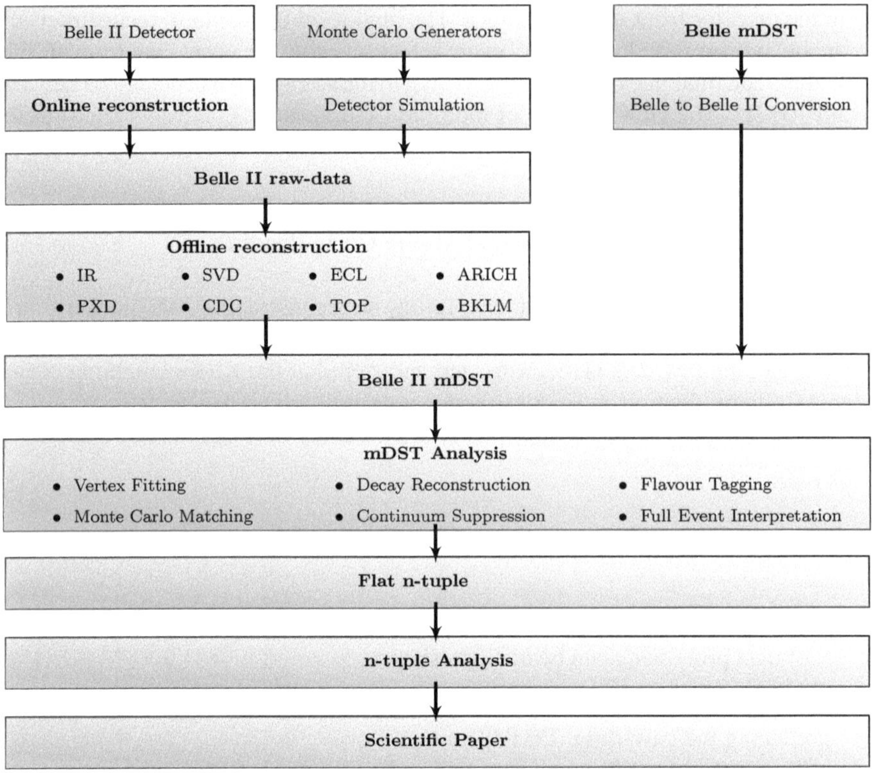

Fig. 2.4 Schematic overview of the data-flow in the Belle II environment. Data is provided by the Belle II detector; the MC generators; or Belle mDST files. `BASF2` is responsible for MC generation, detector simulation, online reconstruction, offline reconstruction, mDST analysis and the Belle to Belle II conversion, as well as writing and reading the different data-formats. Analysis-specific user-code is only required during the n-tuple analysis, which extracts the desired physics observables

2.3.1 Overview

The software responsible for reading in the old `Belle AnalySis Framework` (`BASF`) [9] data-format and representing the data in memory was isolated, cleaned up and compiled into a new library named `belle_legacy`. A new package was introduced in `BASF2` called `b2bii` (Belle to Belle II). It contains three `BASF2` modules developed with the help of the `belle_legacy` library. The conversion process is visualized in Fig. 2.5.

The `B2BIIMdstInput` **module** opens the `PANTHER`-based Belle mDST files and reads the data event-by-event into the main memory. The data of the current event is represented in the memory by a series of `PANTHER` tables.

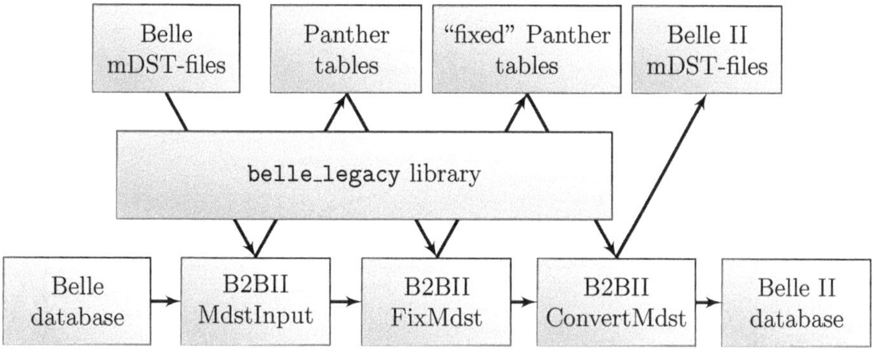

Fig. 2.5 Schematic view of the conversion process of Belle to Belle II mDST files using the BASF2 modules provided by the b2bii package and the original Belle software provided by the belle_legacy library

The B2BIIFixMdst **module** applies various calibration factors onto the PANTHER tables, for instance on the beam-energy, the momenta and error matrices of the fitted tracks, the energy deposition in the ECL, and the particle identification information of the CDC and TOF. It also performs standard cuts to ensure that the selection of the detector data and Monte Carlo events is identical. Finally, π^0 candidates are reconstructed from the γ particle objects and the corrected ECL clusters.

The B2BIIConvertMdst **module** converts the information stored in the Belle PANTHER tables and writes it to the Belle II DataStore. The beam-energy and IP-profile is collected in the BASF2 BeamParameters object and stored in the conditions database of Belle II.

2.3.2 Implementation Details

The detailed matching between PANTHER tables and corresponding BASF2 data-objects is shown in Fig. 2.6. In the following I describe the conversion process in detail for future reference.

2.3.2.1 Event Information

Event information like the beam energy and position of the interaction point are loaded from the Belle conditions database and stored in BeamParameters objects that can be uploaded to the Belle II conditions database.

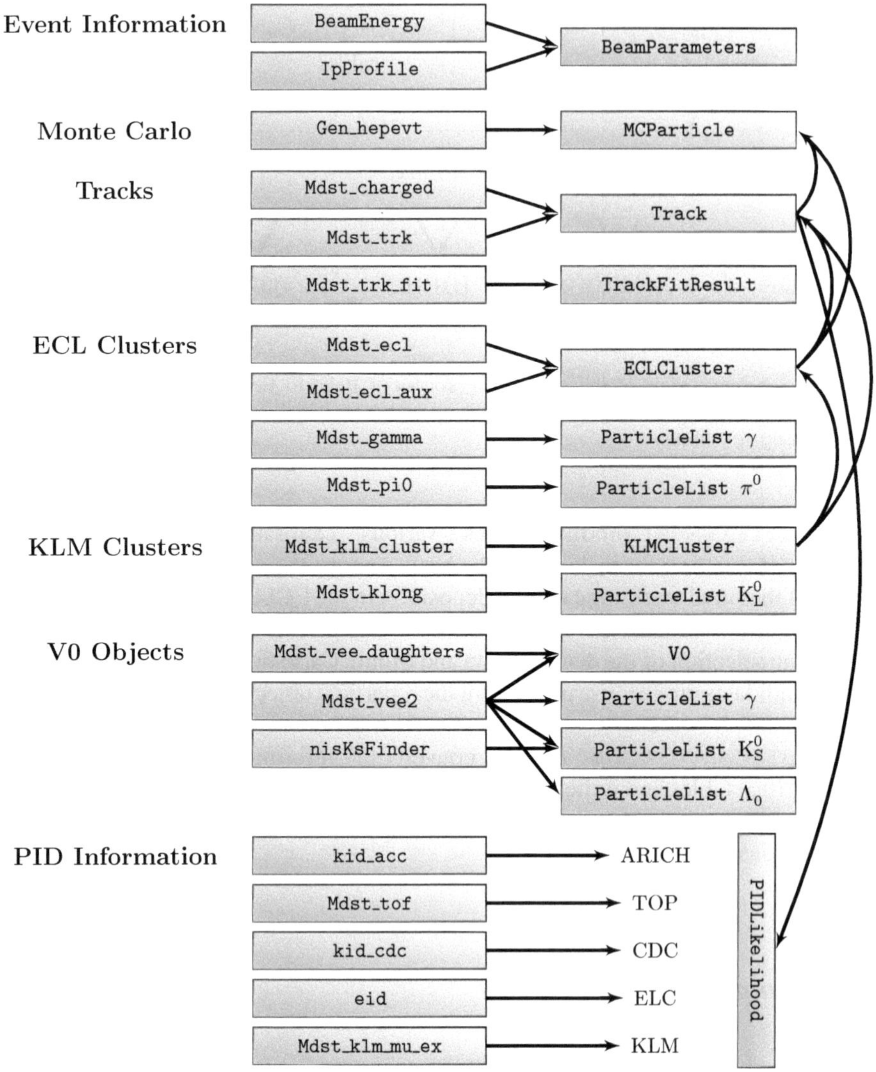

Fig. 2.6 Matching of the Belle PANTHER Tables to the Belle II DataStore-objects and DataStore-relations

BASF assumed a constant magnetic field of 1.5 T even outside the detector region. The conversion accounts for this as well by adopting the same convention in BASF2, where otherwise a 3D map of the magnetic field is used.

2.3.2.2 Monte Carlo

The Monte Carlo information stored in the `Gen_hepevt` table is converted into an `MCParticleGraph`, hence the fine-grained unified Monte Carlo matching algorithm of `BASF2` can be used, and problems contained in algorithms used by `BASF` are avoided [6, Sect. 4.3].

The `Gen_hepevt` table includes special entries for a common mother of beam-background particles (PDG code 911) and for virtual photons (PDG code 0). These entries are ignored during the conversion, because there are no corresponding concepts in Belle II. For instance, in `BASF2` beam-background is indicated by a motherless Monte Carlo particle.

The original Belle software does not provide Monte Carlo information for `KLMClusters`, following the approach of [10, Sect. 5.2] reconstructed K_L^0 are matched to the closest true Monte Carlo K_L^0 within ± 15 degrees in both θ and ϕ.

Furthermore, unlike Belle II Monte Carlo, the Belle Monte Carlo does not provide information on the differentiation between photons generated directly by the fundamental matrix-element calculated by the Monte Carlo generator `evtgen` [11] (hereinafter referred to as gamma) and photons generated afterwards for instance by `PHOTOS` [12] or the simulation [13] (hereinafter referred to as final state radiation) (see [14, Appendix C]). Often a reconstructed particle which misses final state radiation is considered a signal, whereas a missing gamma is considered as a wrong reconstruction. A simple heuristic is applied to distinguish the two cases: Photons from a decay $M \rightarrow AB...\gamma$ are flagged as final state radiation, and photons from a decay $M \rightarrow A\gamma$ are flagged as gammas. In particular photons from $\pi^0 \rightarrow \gamma\gamma$ and $\mathrm{D}^* \rightarrow \mathrm{D}\gamma$ are considered gammas. Other cases like $\mathrm{B} \rightarrow \mu\nu\gamma$ are regarded by the heuristic as final state radiation and have to be treated by the analyst themselves.[7]

The official Belle Monte Carlo campaigns produced ten times the real integrated luminosity in $\mathrm{b\bar{b}}$ events and six times that in continuum events, however some inconsistencies were encountered during the development of the conversion software, which were fixed if possible: The Monte Carlo campaign deleted the 8 left-most bits of the 32 bit long PDG codes during the Monte Carlo simulation.[8] During the conversion these corrupted PDG codes are restored by matching their lower 24 bit to known PDG codes. In the official Belle Monte Carlo campaign from 2010 for $\mathrm{B} \rightarrow \mathrm{u}\ell\nu$ and other rare B decays, the mass of almost all MC particles is set to zero, which can lead to wrong results if this quantity is used during the analysis. However, this information is redundant since the correct mass of the MC particles can be calculated using either the PDG values or the MC four-momenta.

---

[7] In this case, photons from initial and final state radiation are physically indistinguishable, since the corresponding amplitudes interfere. Actually, there is no correct answer to the question of whether the photon is final state radiation or not. Hence, the behavior of the heuristic is not wrong, but probably unexpected by the analyst, because the initial state radiation amplitude dominates in this decay.

[8] `BASF` already implemented a function for recovering the lost bits, but it was apparently not applied.

2.3.2.3 Tracks

The tracking output of BASF is transformed and stored into Track and associated TrackFitResult objects. The transformation is unique but non-trivial because the two experiments employ different track parameterizations and conventions for the reference point of the track.

2.3.2.4 ECL Clusters

The ECL information is stored in the ECLCluster object and two ParticleLists are filled containing the γ and π^0 candidates created by B2BIIFixMdst earlier. The lists are named gamma:mdst and pi0:mdst, respectively.

2.3.2.5 KLM Clusters

The KLM information is stored in the KLMCluster object and a ParticleList is filled containing K^0_L candidates. The list is named K_L0:mdst.

2.3.2.6 V0 Objects

The output of the V0 Finder is directly transformed into ParticleList objects containing candidates for γ, K^0_s and Λ. The lists are named gamma:v0mdst, K_S0:mdst and Lambda0:mdst, respectively. Additional quality information is stored in the ExtraInfo field of the Particle objects under the keys goodKs, ksnbVLike, ksnbNoLam and ksnbStandard.

2.3.2.7 PID Information

The PID information provided by the different Belle sub-detectors is mapped to similar Belle II sub-detectors, so that the physical meaning of the information is partially preserved. In particular the Belle time-of-flight (TOF) and Aerogel Cherenkov counter (ACC) detectors are mapped to the Belle II time-of-propagation (TOP) and Aerogel ring imaging Cherenkov (ARICH) detectors, respectively.

2.3.2.8 Relations

Finally, some of the created data-objects are related to one another (see Fig. 2.6). Hence, BASF2 relations are created: from the ECLCluster to the MCParticle and Track which are responsible for the creation of the cluster; similarly from the KLMCluster to the ECLCluster and Track; from the Track to the MCParticle that created it; and from the Track to the corresponding PIDLikelihoods. Additional relations are created between the Particle objects in the created ParticleLists and the corresponding MCParticle and PIDLikelihoods. The links between TrackFitResults to Tracks; Tracks to Particles; and Clusters to Particles are not represented by relations in BASF2.

2.3.3 Validation

In order to ensure the correctness of the conversion, a study was performed with Monte Carlo simulated Belle events. For each of the following six physical processes 100,000 simulated events were investigated, partitioned according to the relative integrated luminosity of all relevant experiments 7–65:

- generic $\Upsilon(4S)$ events:
  - $\Upsilon(4S) \to B^+B^-$ (evtgen-charged),
  - $\Upsilon(4S) \to B^0\bar{B}^0$ (evtgen-mixed);
- continuum events:
  - $e^+e^- \to c\bar{c}$ (evtgen-charm),
  - $e^+e^- \to q\bar{q}$ (uds) (evtgen-uds),
  - $e^+e^- \to \pi^+\pi^-$ (qed-tautau);
- and signal events $B^+ \to \pi^+\nu_\tau$ (signal).

Furthermore, for each of the following beam conditions 100,000 events recorded by the Belle experiment were investigated, partitioned according to the relative integrated luminosity of all relevant experiments 7–65:

- on-resonance events recorded at the center of mass energy of the $\Upsilon(4S)$ resonance;
- and off-resonance events recorded 60 MeV below.

The events were processed with the old BASF framework and more than 360 quantities[9] were extracted from the PANTHER tables shown in Fig. 2.6. The complete list of extracted quantities can be found in Sect. A.1. Afterwards the events were processed a second time with the new BASF2 software using the b2bii conversion, and the same quantities were extracted.

2.3.3.1 Observed Differences

Most quantities do not differ.

[9]For instance: Kinematic quantities like four-momenta, Monte Carlo information, PID information and beam-parameters.

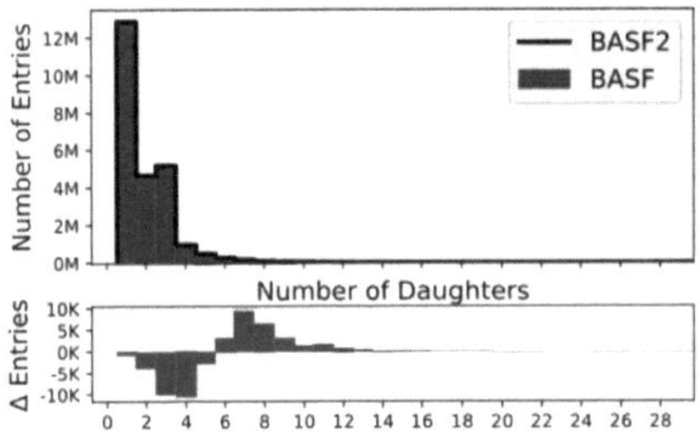

(a) Example of major differences: The number of daughters of the Monte Carlo particle objects is shifted to smaller values because virtual photons are ignored during the conversion.

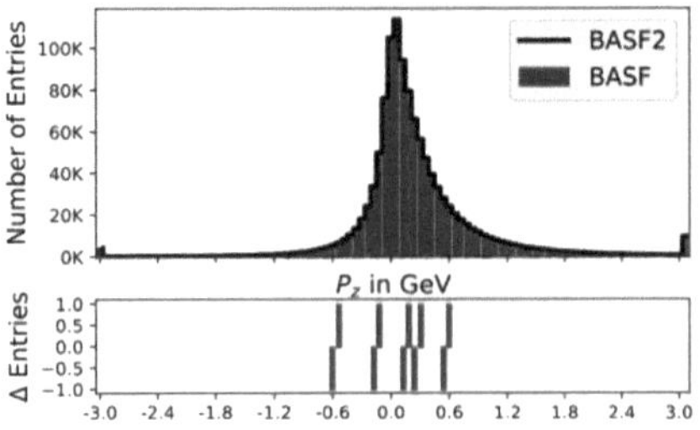

(b) Example of minor differences: Momentum of Tracks in z direction exhibiting migration during the conversion due to numerical imprecision and special floating point values.

Fig. 2.7 Comparison of `BASF` (Belle) and `b2bii` (Belle II). The leftmost (rightmost) bin represents the underflow (overflow) bin. The upper plots show the superimposed Belle and Belle II Monitoring Histograms (the total number of entries is shown as a black line). The lower plots show the differences between Belle and Belle II, hence a positive (negative) difference means there are less (more) entries in the Belle II Monitoring Histogram

The observed differences between `BASF` and `b2bii` were further investigated and either corrected or classified as harmless. Minor differences occur due to small shifts caused by numerical imprecision leading to the migration of events between adjacent bins, especially for values near zero, and differences in the treatment of special floating point values such as infinity and NaN (Not a Number) leading to migration from the overflow/underflow bin to the bin including zero in rare cases (see Fig. 2.7b).

Further differences are found: in the PDG codes of `MCParticles` due to the recovery of the full 32 bit as mentioned above; the number of daughters of the `MCParticles` due to the unconverted virtual photons occurring in nuclear interactions between the hadronic final state particles and the detector material (see Fig. 2.7a); and in all kinematic quantities of $V0$ and π^0 objects after the mass-constrained vertex fit caused by different software employed to fit the vertices.

2.3.3.2 Experiment Dependency

The changes in hardware and software between the 31 Belle experiments influence the quantity and quality of the recorded data. In order to carry out physics analyses and compare them with studies on Belle II Monte Carlo events, quantities should be used, which are robust against differences between the Belle experiments. The beam-induced background in Belle II is expected to be increased by a factor 20–30. Quantities that are already fluctuating between Belle experiments are likely to be highly dependent on the assumed background conditions in Belle II. One of the most

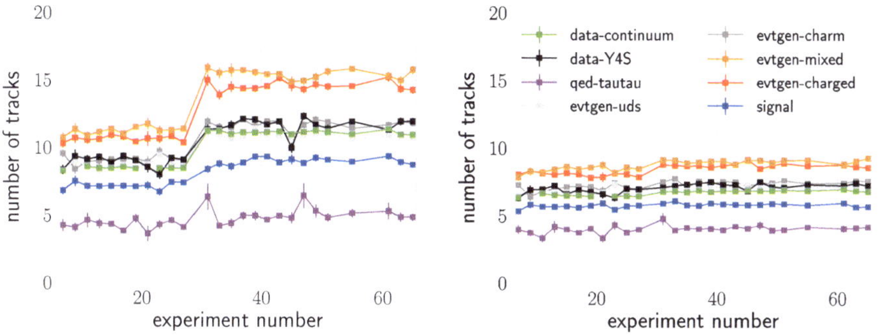

(a) All tracks found by the tracking software.

(b) Only good tracks: Impact parameters less than 2 cm and 4 cm in transverse and beam axis direction, respectively.

Fig. 2.8 The mean number of reconstructed tracks over the Belle experiments. The error bars show the uncertainty on the mean value, however they are mostly too small to be visible. The increase in the number of tracks after experiment 27 is caused by the switch to the improved reconstruction software and the replacement of the SVD

basic properties of an event is the number of fitted tracks. As can be seen in Fig. 2.8 there are large differences between the Belle experiments for this quantity.

Therefore, the following selections are applied to Belle events in the remainder of this work: **good tracks** are selected with impact parameters less than 2 and 4 cm in transverse and beam axis direction, respectively (see Fig. 2.8); and **good clusters** are selected without an associated track and with an energy of 50, 100, 150 MeV in the barrel, forward and backward region, respectively (see Fig. 2.9). These are the same cuts which were usually applied in Belle analyses.

Some differences between the experiments remain after the cuts. In particular due to the switch to the improved reconstruction software and the replacement of the SVD after experiment 27.

2.3.3.3 Monte Carlo Simulation and Detector Data Differences

This thesis uses the last official Belle Monte Carlo campaign (prefixed with `evtgen-`). In addition the latest special Monte Carlo campaign for rare B decays is used (prefixed with `special-`). And finally some dedicated Monte Carlo samples for QED background were investigated (prefixed with `qed-`). Those were usually not used in Belle B meson analyses, because the `hadronBJ` skim suppresses them very efficiently.

Table 2.3 shows the extracted fraction and the corresponding fractions in the 7th Belle II Monte Carlo campaign for comparison.

The simulation of continuum events, that is events which do not contain a $\Upsilon(4S)$, is challenging due to the required knowledge of the correct QCD hadronization

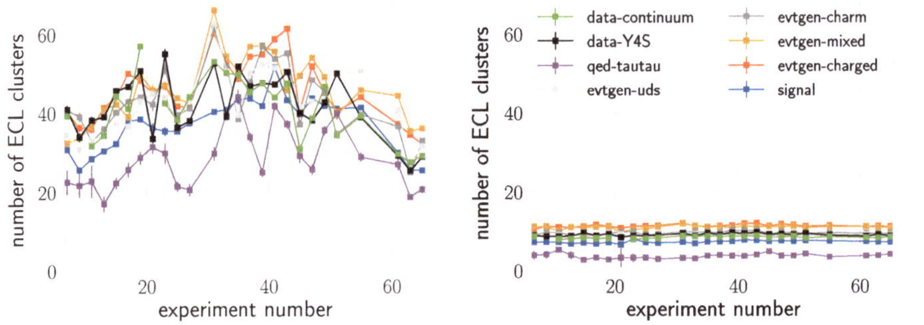

(a) All clusters found by the reconstruction software.

(b) Only good clusters: No associated tracks and E < 50, 100, 150 MeV in the barrel, forward and backward region, respectively.

Fig. 2.9 The mean number of reconstructed clusters over the Belle experiments. The error bars show the uncertainty on the mean value, however they are mostly too small to be visible. The large differences between the experiments are due to different beam conditions

Table 2.3 Number of simulated events corresponding to the recorded luminosity of $0.711\,\mathrm{ab}^{-1}$ normalized to the number of the recorded events at the $\Upsilon(4S)$ resonance. Belle simulated rare and $b \to u\ell\nu$ decays separately, while Belle II includes them in the evtgen-charged and evtgen-mixed samples. In contrast to the official Belle MC campaign, the relative ratio of $B^+B^-/B^0\bar{B}^0 = 0.514/0.486$ was taken into account

Type	Belle	Belle II
evtgen-charged	0.1264	0.1310
evtgen-mixed	0.1195	0.1238
evtgen-charm	0.2933	0.3078
evtgen-uds	0.4716	0.5534
qed-tautau	0.1996	0.2052
qed-mumu	0.2395	–
qed-eemm	0.2539	–
special-mixedrare	0.00102	–
special-chargedrare	0.000824	–
special-mixedulnu	0.000859	–
special-chargedulnu	0.000923	–

factors and branching fractions. In addition, the exact composition of the continuum background is poorly documented and not all components are part of the available official (or even unofficial) Monte Carlo campaign, e.g. $e^-e^+ \to \gamma\gamma$, is not available.

In consequence, the continuum description of Belle is not in agreement with the observed data. The off-resonance data recorded by the Belle detector can be used as a cross-check and sometimes replacement for Monte Carlo simulated continuum events. In a physics analysis the observables are usually extracted using a model for

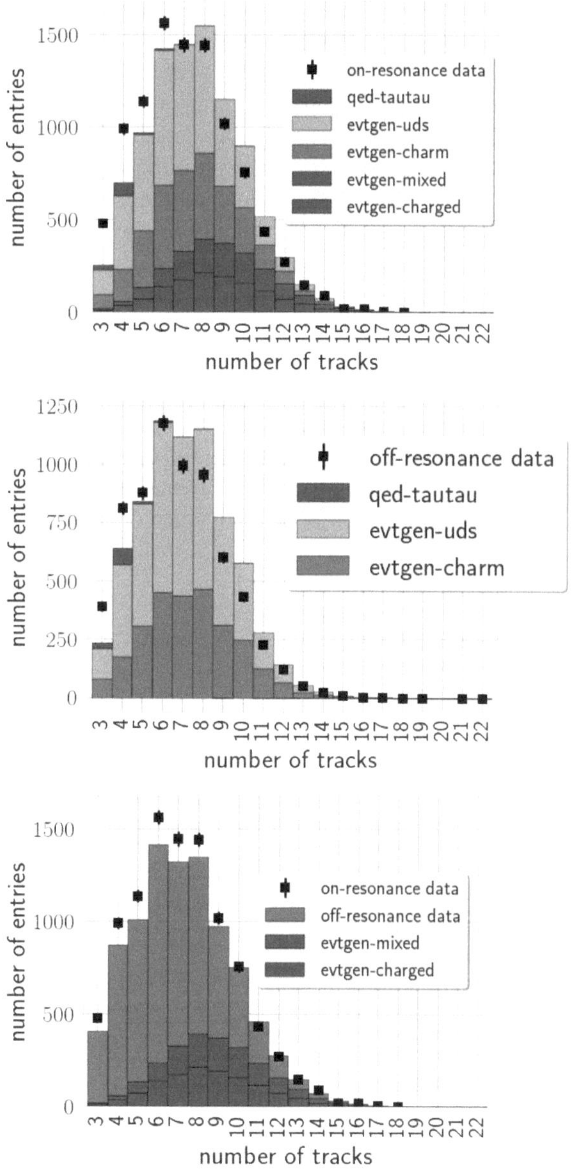

Fig. 2.10 The distribution of the number of good tracks in each event

the signal and background components, the relative fractions are usually not fixed and determined directly from data. Nevertheless, the continuum component description is often a leading systematic uncertainty in Belle analyses [15].

The continuum composition was investigated using 10,000 simulated and recorded events for each component, scaled according to the fractions stated in Table 2.3 of the considered QED backgrounds only $e^-e^+ \to \tau^-\tau^+$ contributes. The other components are suppressed by the so-called `hadronBJ` cut, which is applied per default for all B analyses. As can be seen from Fig. 2.10 the Monte Carlo continuum description is incomplete, it underestimates (overestimates) the events in the low-multiplicity (high-multiplicity) region. In contrast, the $\Upsilon(4S)$ Monte Carlo simulation combined with the off-resonance data fits the data better.

The over-estimation of continuum background, most-likely the $c\bar{c}$ component, leads to an increase of D meson candidates and finally to more combinatorial background for B mesons. An example is shown in Fig. 3.15, where D mesons were reconstructed on Monte Carlo simulated events, on-resonance data and off-resonance data.

Throughout this thesis, these expected differences between Monte Carlo and data are observed in several places. Section 3.3.3 describes techniques to mitigate the influence of these differences in the context of multivariate classification. Section 4.3.2 investigates the continuum component in the context of exclusive tagging. In the benchmark analysis in Chap. 5, the continuum component is a large background in the hadronic τ decay-channels. Hence, the continuum description was investigated in detail and compared to off-resonance data.

2.4 Conclusion

The Belle to Belle II Conversion enables Belle II physicists to analyze the dataset recorded by Belle using `BASF2`. The conversion process was validated on a basic level by ensuring the same output for a large number of quantities. Differences which emerged were studied and explained.

In order to validate `BASF2` on a global level, physics analyses have to be performed and compared to results published by the Belle collaboration. In this thesis the decay $B \to \tau\nu_\tau$ is investigated and compared to [15, 16].

Other measurements using the `b2bii` conversion are in preparation:

- the branching ratio of $B^+ \to \ell^-\nu\gamma$ with hadronic tagging [17],
- the branching ratio of $B^+ \to \ell^-\nu\gamma$ with semileptonic tagging [18],
- the branching ratio of $B^+ \to \ell^-\nu$ with inclusive tagging,
- and the search for $B^0_s \to \phi\pi^0$ [19].

Furthermore, `b2bii` can be used to study the performance differences between the `Belle` and `Bellef II` experiment, to optimize the latter as soon as first data has been collected.

Finally, the conversion ensures the preservation of the legacy of the Belle experiment: The full recorded dataset of nearly 1 ab^{-1} of data, which led mankind to the verification of the CKM mechanism and the observation of tetra-quarks.

References

1. A. J. Bevan et al., The physics of the B factories. Eur. Phys. J. C **74**, 3026 (2014). https://doi.org/10.1140/epjc/s10052-014-3026-9
2. A. Abashian, K. Gotow, N. Morgan, L. Piilonen, The Belle detector. Nucl. Instrum. Methods Phys. Res. Sect. A: Accel. Spectrom. Detect. Assoc. Equip. **479**(1), 117–232 (2002). https://doi.org/10.1016/S0168-9002(01)02013-7
3. T. Keck, The Full Event Interpretation for Belle II. M.A. thesis. KIT (2014), https://ekp-invenio.physik.uni-karlsruhe.de/record/48602
4. Z. Doležal, S. Uno, Belle II Technical Design Report. Technical report KEK (2010). arXiv: 1011.0352 [hep-ex]
5. C. Schwanda, SuperKEKB machine and Belle II detector status. Nucl. Phys. B—Proc. Suppl. **209**(1), 70–72 (2010). https://doi.org/10.1016/j.nuclphysbps.2010.12.012
6. C. Pulvermacher, Analysis Software and Full Event Interpretation for the Belle II Experiment. Ph.D. thesis. KIT, (2015), https://ekp-invenio.physik.uni-karlsruhe.de/record/48741
7. http://www-superkekb.kek.jp/documents/luminosityProjection_160802.pdf. Accessed 20 Jan 2017
8. D. Kim, The software library of the Belle II experiment, in *Nuclear and Particle Physics Proceedings*, vol. 273 (2016), pp. 957–962. https://doi.org/10.1016/j.nuclphysbps.2015.09.149
9. R. Itoh, BASF–BELLE AnalysiS Framework, in *Proceedings, 9th International Conference on Computing in High-Energy Physics (CHEP 1997)* (1997), http://www.ifh.de/CHEP97/paper/244.ps
10. K. Hara, Calibration of low momentum K_L^0 efficiency for veto usage in missing E analyses. Belle Note 1228 (2012)
11. D.J. Lange, The EvtGen particle decay simulation package. Nucl. Instrum. Methods A **462** (2001). https://doi.org/10.1016/S0168-9002(01)00089-4
12. E. Barberio, B. van Eijk, Z. Was, Photos—a universal Monte Carlo for QED radiative corrections in decays. Comput. Phys. Commun. **66**(1) (1991). https://doi.org/10.1016/0010-4655(91)90012-A
13. S. Agostinelli et al., GEANT4: a simulation toolkit. Nucl. Instrum. Methods **A506** (2003). https://doi.org/10.1016/S0168-9002(03)01368-8
14. A. Ryd et al., EvtGen A Monte Carlo Generator for B—Physics. User manual, http://evtgen.warwick.ac.uk/static/docs/EvtGenGuide.pdf
15. B. Kronenbitter et al., Measurement of the branching fraction of $B^+ \to \tau^+\nu_\tau$ decays with the semileptonic tagging method. Phys. Rev. D **92**(5), 051102 (2015). https://doi.org/10.1103/PhysRevD.92.051102
16. K. Hara, Y. Horii, T. Iijima, Evidence for $B^- \to \tau^-\bar{\nu}_\tau$ with a hadronic tagging method using the full data sample of Belle. Phys. Rev. Lett. **110** (2013). https://doi.org/10.1103/PhysRevLett.110.131801
17. F. Metzner, Analysis of $B^+ \to \ell^+\nu\ell\gamma$ decays with the Belle II Analysis Software Framework. M.A. thesis. KIT (2016), https://ekp-invenio.physik.uni-karlsruhe.de/record/48845
18. M. Gelb, Search for the decay $B^+ \to l^+\nu\gamma$ with the Full Event Interpretation. Belle Note 1474 (2017)
19. F. Fichter, Search for $B_s \to \phi\pi^0$ Decays at the Belle Experiment. M.A. thesis. KIT (2016), https://ekp-invenio.physik.uni-karlsruhe.de/record/48880.26

Chapter 3
Multivariate Analysis Algorithms

In recent years, the field of multivariate analysis and machine learning evolved rapidly, and provided powerful techniques, which are currently adopted in all fields of science. Prominent use-cases include: image and speech recognition, stock market trading, fraud detection, and medical diagnosis.

In high energy physics (HEP) multivariate analysis (MVA) methods are extensively used: to identify interesting collision in the trigger system; reject beam-induced hits in the drift chamber during track finding; infer the deposited energy in the calorimeter; provide particle identification information; reject particle candidates from combinatoric and physics background in an analysis.

The Belle II collaboration decided to provide a multivariate analysis package named `mva` as part of `BASF2` to encourage physicist to take advantage of existing methods and to keep up with the recent developments in the field. The package provides: a common code base for `BASF2`; support for all major MVA frameworks; transparent run-dependent loading of the fitted models from the `Belle II Conditions Database`; tools for automating fitting, inference and evaluation of models.

The `mva` package and many algorithms based on it were implemented during this thesis. In the following chapter I outline the theoretical foundation of multivariate analysis and machine learning (Sect. 3.1), describe technical aspects of the `mva` package (Sect. 3.2), and present applications based on the package (Sect. 3.3).

3.1 Theory

Multivariate Analysis (MVA) algorithms are used to describe the distribution of observed data $\vec{x}$ and to deduce properties of the underlying process generating the data; the algorithms are based on multivariate statistics. Common tasks include: the approximation of a function $f(\vec{x})$ (regression Sect. 3.1.1); the separation of data-points into signal and background (classification Sect. 3.1.2); and grouping of similar data-points (clustering).

T. Keck, *Machine Learning at the Belle II Experiment*, Springer Theses,
https://doi.org/10.1007/978-3-319-98249-6_3

Machine learning (ML) provides an effective way to automatically learn the required statistical model using an appropriate domain-specific dataset, i.e. knowledge is extracted from experience, which can be viewed as a simple form of artificial intelligence. Usually, machine learning algorithms include a fitting-phase during which a statistical model is learned from the provided (training) dataset, and an inference-phase during which the statistical model is used to infer the desired (target) information for a new independent (test) dataset.

Depending on the available feedback during the fitting-phase one can distinguish between three different types of learning: the target information is provided during the fitting-phase (supervised learning); no additional information is provided during the fitting-phase (unsupervised learning); rewards and punishments are provided in a dynamic environment (reinforcement learning).

In HEP the target information is usually known from Monte Carlo simulated events, or can be inferred on data via data-driven techniques (see Sect. 3.3.3). Therefore, most tasks can be solved using supervised learning. In supervised learning the target information $y \sim f(\vec{x}) + \epsilon$ is predicted by a statistical model $\widehat{f}$ using a feature vector $\vec{x}$. The statistical model has internal degrees of freedom (so-called weights $\vec{w}$), which can be adapted by an algorithm to minimize the discrepancy of the true value y from the predicted value $\widehat{y} = \widehat{f}(\vec{x}, \vec{w})$, where the discrepancy is defined by a loss-function $\mathcal{L}(y, \widehat{y}, \vec{w})$.

Detailed introductions into the topic can be found in [1, 2]. The relevant concepts are summarized in the remainder of this section.

3.1.1 Regression

A statistical model $\widehat{f}(\vec{x}, \vec{w})$, which estimates the value of a stochastic function $y = f(\vec{x}) + \epsilon$ is called a regressor. It can be interpreted as the function which minimizes the risk functional

$$R(\vec{w}) = \int \mathcal{L}(y, \widehat{f}(\vec{x}, \vec{w}), \vec{w}) \mathrm{d}P(\vec{x}, y), \tag{3.1}$$

where $P(\vec{x}, y)$ is the unknown joint probability distribution [3]. In the framework of empirical risk minimization (ERM) the risk functional is approximated by the empirical risk using a training dataset $(\vec{x}_i, y_i)$

$$R_{\text{emp}}(\vec{w}) = \sum_i \mathcal{L}(y_i, \widehat{f}(\vec{x}_i, \vec{w}), \vec{w}). \tag{3.2}$$

The weights $\vec{w}$ of the statistical model are chosen so that they minimize the empirical risk. This is known as the principle of empirical risk minimization [3].

A suitable loss function can be chosen using the ERM framework [4] or based on the maximum likelihood principle. Typical examples are: the squared loss $(\widehat{y} - y)^2$;

and the absolute loss $|\widehat{y} - y|$. Often additional terms are added to the loss function which depends on the weights $\vec{w}$ of the model, to impose certain smoothness and regularization assumptions on $\widehat{f}$.

A well-known example for the application of the principle of ERM is the linear regression model $y_i = \vec{x}_i \cdot \vec{w}$ with a squared loss, which leads to the well known least-square solution for the weights of the linear regression model $\vec{w} = (X^T X)^{-1} X^T \vec{y}$, where $X = (\vec{x}_2, \vec{x}_2, \ldots, \vec{x}_n)$ and $\vec{y} = (y_1, y_2, \ldots, y_n)$.

Various methods for regressions have been developed. One can distinguish between:

- **Parametric methods** with a fixed parametrized model structure and a finite number of unknown parameters that are estimated from the training dataset. Typical examples include: Generalized Linear Models [5] and Artificial Neural Networks [1, Chap. 5], [2, Chap. 11].
- **Nonparametric methods** estimate an appropriate model structure from the training dataset. Typical examples include: Kernel Density Estimator [6], [2, Chap. 6] and Decision Trees [2, Chap. 9].

As stated above, the estimated function f is stochastic. Therefore, its value y is a stochastic variable, which is distributed according to a conditional probability density function $P(y|\vec{x})$. In consequence, regression can be generalized to predict location, scale and shape parameters using a distribution assumption [7]; or the conditional density function itself without assuming a particular distribution using quantile regression [8] or mixture density networks [1, Sect. 5.6].

3.1.2 *Classification*

A statistical model $\widehat{f}$, which distinguishes signal ($y = 1$) from background ($y = 0$) is called a classifier. Classification can be viewed as a particular form of regression. The classifier is a regressor for the signal probability $P(y = 1|\vec{x})$. At a certain threshold (working point) C, data-points with $\widehat{f} > C$ are classified as signal, and background otherwise. The choice of the working point depends on the expected cost of signal classified as background (type I error) and background classified as signal (type II error), or equivalently on the desired signal efficiency and purity (which are usually more closely related to the figure of merit used in physics analyses).

There is a close connection to the theory of hypothesis testing, the statistical model can be interpreted as a test-statistic $T = \widehat{f}(\vec{x}, \vec{w})$. The most efficient test statistic at a given significance level to distinguish between two simple hypotheses is given by the Neyman-Pearson Lemma

$$T_{\mathrm{NP}}(\vec{x}) = \frac{\mathrm{PDF_S}(\vec{x})}{\mathrm{PDF_B}(\vec{x})}, \tag{3.3}$$

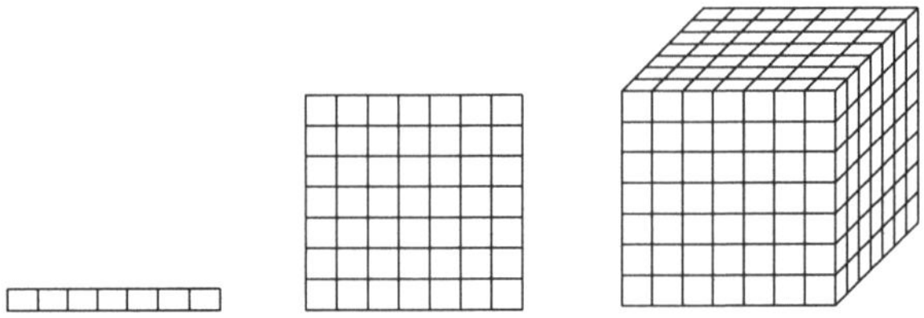

Fig. 3.1 Visualization of the curse of dimensionality. The phase-space of an N dimensional function grows exponentially, thus the number of events necessary to sample a high-dimensional probability density function grows exponentially as well and is unviable

where PDF denotes the probability density functions (PDF) of signal and background data-points. A sound introduction to hypothesis testing and the Neyman-Pearson Lemma can be found in the original paper [9].

The PDFs are usually unknown and cannot be sampled in a high dimensional feature space $\vec{x}$, due to the curse of dimensionality (see Fig. 3.1), in consequence T_{NP} is unknown and has to be approximated with analytical (using distribution assumptions) or numerical (using machine learning) methods. Two approaches can be distinguished:

- **Generative Methods** like Fisher's discriminant [10], Kernel Density Estimators [6], [2, Chap. 6] or Gaussian mixture models [1, Chap. 9]; which approximate the probability density functions for signal $\widehat{f_S} \approx \mathrm{PDF_S}(\vec{x})$ and background $\widehat{f_B} \approx \mathrm{PDF_B}(\vec{x})$ separately, and can therefore be used to generate new data-points.
- **Discriminative Methods** like Boosted Decision Trees (BDT) [11], [2, Chap. 10], Support Vector Machines (SVM) [1, Chap. 7], [2, Chap. 12] or Artificial Neural Networks (ANN) [1, Chap. 5], [2, Chap. 11]; which directly approximate $\widehat{f}(\vec{x}) \approx T_{\mathrm{NP}}$, usually under the assumption that the dimension d of the hyperplane separating (discriminating) signal and background in the feature space is much smaller than the dimension D of the feature-space itself $d \ll D$.

The methods differ in: the convergence rate against the optimal solution with increasing statistics, their robustness against outliers and measurement errors; the bias and variance of the obtained model (see Sect. 3.1.3).

Typical loss functions used for classification are: the logistic loss [4]; the hinge loss [4]; and the cross entropy [1, Sect. 4.3.2]. The loss-functions differ in their robustness against outliers and their convergence rate (see [4]).

3.1.3 Model Complexity

The complexity or number of degrees of freedom (NDF) of the statistical model strongly influences the prediction error of the model on an independent test dataset.

3.1.3.1 Bias-Variance Dilemma

There are three possible sources of errors in any statistical model $\widehat{f}(\vec{x})$ describing a multivariate distribution $y \sim f(\vec{x}) + \epsilon$.

1. (**Bias**) The model is not complex enough to describe all relevant aspects of the data—the model is under-fitted. This leads to a biased model $\text{Bias}(\widehat{f}) = \text{E}(\widehat{f} - f)$.
2. (**Variance**) The model is too complex and is dominated by the statistical fluctuations of the training data—the model is over-fitted. This leads to a model with a large variance $\text{Var}(\widehat{f}) = \text{E}(\widehat{f}^2) - \text{E}(\widehat{f})^2$.
3. (**Irreducible**) The intrinsic noise ϵ of the underlying distribution introduces an irreducible error $\text{Var}(y) = \text{Var}(\epsilon)$.

Mathematical we can expand the expected quadratic deviation of the statistical model $\widehat{f}$ from the true value y into these three components

$$\text{E}\left[(y - \widehat{f}(\vec{x}))^2\right] = \text{Bias}\left[\widehat{f}(\vec{x})\right]^2 + \text{Var}\left[\widehat{f}(\vec{x})\right] + \text{Var}\left[y\right].$$

Figure 3.2 visualizes the trade-off between bias and variance as a function of the model complexity. In the literature this fact is known as the Bias-Variance Dilemma [1, Sect. 3.2], [2, Sect. 2.9]. Since the over-fitting effect cannot be seen on the training data, it is crucial to always test the model on an independent validation dataset. Choosing the optimal model complexity is the key to the successful employment of MVA methods.

3.1.3.2 Controlling the Model Complexity

The size of the training dataset at hand defines an upper bound for the feasible model complexity. The model cannot resolve the underlying distribution f below the threshold given by statistical fluctuations, i.e. the bigger the training dataset, the more degrees of freedom of a model can be constrained.

MVA algorithms usually provide several hyper-parameters Θ (in contrast to the parameters/weights of the statistical model itself) to control the complexity of the

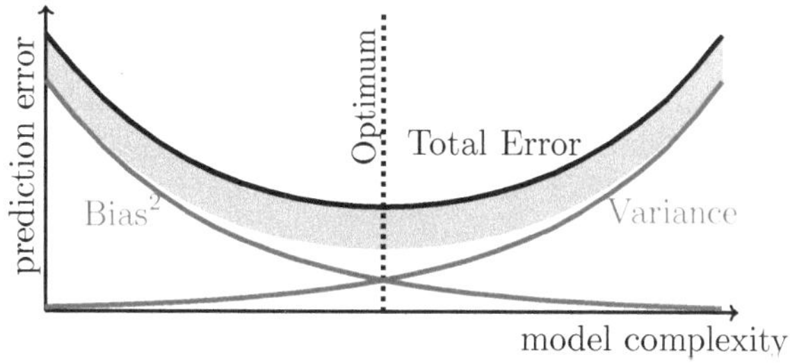

Fig. 3.2 Visualization of the trade-off between bias and variance of a MVA method depending on the model complexity

underlying model. The theoretical foundation is provided by the principle of structural risk minimization (SRM) [3]. Following [3], one can distinguish three different concepts of SRM:

- **Structure given by the architecture**—Hyper-parameters which define the architecture of the model directly control the global NDF of the model. Typical examples are: the number of hidden layers and neurons in an artificial neural network; the number of trees and their depth in a boosted decision tree.
- **Structure given by the learning procedure**—Hyper-parameters which control the effective NDF in different regions of the feature-space during the fitting-phase. Typical examples are: weight-decay in an artificial neural network; pruning algorithms for a decision tree; and boosting algorithm in ensemble methods.
- **Structure given by preprocessing**—Hyper-parameters which control the degree of degeneracy in the input data by applying a transformation to the input features prior to the fitting-phase. Typical examples are: binning and smoothing of the input features; dimensionality reduction using principal component analysis.

SRM provides another interpretation of the Bias-Variance Dilemma introduced above, where the bias is given by the empirical risk and the variance is introduced by the generalization error due to the finite sample size. Using the hyper-parameters Θ, a series of models $\widehat{f_1}, \widehat{f_2}, \ldots, \widehat{f_n}$ with increasing complexity (defined by their Vapnik–Chervonenkis (VC) dimension h [2, Sect. 7.9]) is defined. The empirical risk is minimized for each model (meaning the model is fitted). Using the framework of SRM, an upper bound $C(h, l)$ for the generalization error introduced by the finite training sample size can be calculated, i.e, an upper bound for the difference $|R - R_{\mathrm{emp}}|$ between the actual risk R and the empirical risk R_{emp}. The principal of structural risk minimization defines the optimal model as the one with the minimal guaranteed risk $R_{\mathrm{emp}} + C$.

However, the upper bound C can only be calculated for certain models, and is usually only a weak bound. In practice, the optimal model is selected by optimizing the associated hyper-parameters, e.g. using grid or random-search [12]; gradient descent algorithms [13]; or Bayesian optimization algorithms [14].

3.1.4 Data Analysis in High Energy Physics

Data analysis plays a crucial role in HEP experiments like Belle II. There are unique characteristics and associated challenges.

- The inherent stochastic nature of quantum field theory rules out a ground truth for individual data-points. Therefore, supervised machine learning and data analysis have to rely on Monte Carlo simulation or on advanced data-driven techniques to establish a ground truth.
- The investigated phenomena are usually very subtle and cause only small effects. Therefore, large amounts of events (often in the order of billions) have to be analyzed in a high-dimensional feature-space.

- The statistical and systematic uncertainties have to be treated very carefully to produce sound scientific results.

Consequently, the HEP community developed data analysis strategies suited for these requirements. Following the history of data analysis in HEP, I describe the advantages and disadvantages of the traditional cut-based (Sect. 3.1.4.1), the current multivariate (Sect. 3.1.4.2) and the prospective deep learning (Sect. 3.1.4.3) approach. An in-depth introduction can be found in [15].

3.1.4.1 Traditional Cut-Based Analyses

In the traditional cut-based approach, the physicist determines a series of one-dimensional cuts on the dataset, to incrementally increase the signal-to-noise ratio in the signal-region.

For instance, one could cut on the K^+ PID information, the invariant mass of the π^0, and the χ^2 probability of the vertex fit to increase the signal-to-noise ratio in the reconstruction of the decay $D^0 \rightarrow K^-\pi^+\pi^0$.

The cuts are determined on Monte Carlo simulated events or in control regions.[1] The final signal region is only unblinded once the whole analysis strategy is fixed.

Physics knowledge can be easily incorporated into this process: phase-space with large systematic uncertainties (for instance due to unreliable Monte Carlo simulation or threshold effects in reconstruction algorithms) can be excluded; the dataset is pre-processed and reduced to a suitable subset for this analysis (for instance intermediate particles are reconstructed and irrelevant features are dropped); new physically motivated features can be constructed (for instance angles in particular reference frames); and erroneous assumptions or software can be spotted by experienced analysts.

It is relatively easy to estimate the systematic uncertainty introduced by the cuts, by varying the cut in a certain range. Although, this common procedure has been criticized (see [15, Sect. 8.4.3]). Moreover, the agreement between Monte Carlo simulation and data can be verified after each cut.

The fixed analysis procedure is usually simple enough to be included in recasting frameworks, which recast existing analysis to provide bounds for new physics scenarios, by automatically analyzing custom signal Monte Carlo simulated events [16].

On the other hand, multivariate correlations are not (or only to a small degree) taken into account, which leads to reduced signal efficiency. Possible correlation between the cuts are often neglected. There is no obvious choice for the range in which the cuts are varied to estimate the systematic uncertainties. Finally, the process of manual cut determination and feature engineering is time-consuming and error-prone in itself.

[1] A dataset without signal, but comparable background properties.

3.1.4.2 Multivariate Analysis

In a multivariate analysis approach, the physicist uses MVA methods like machine learning to create a statistical model of the dataset. Either this can be used to infer the signal probability of each event, thereby increasing the signal-to-noise ratio in the signal-region, or to directly model and extract the signal component in the dataset. The statistical model is determined on Monte Carlo simulated events, in control regions or using other data-driven methods. Sometimes it is possible to calibrate the model on data (see Sect. 3.3.1.2).

The multivariate correlations between the input features are taken into account, frequently this leads to an improved signal efficiency by a factor two or more (e.g. [17]). The available dataset is therefore exploited more efficiently and the statistical uncertainty of the final physics result is reduced.

The preprocessing of the dataset and the creation of useful physically motivated high-level features is still handled by the physicist. The subsequent model building is automatized and can be applied to similar classes of analyses without further human intervention (see Sect. 3.3.1.1).

On the other hand, the validity of the statistical model has to be carefully verified using control regions and channels, to avoid possible over-fitting and mis-modeling effects. Physics knowledge must still be incorporated into this process to exclude problematic phase-spaces. Due to the increased complexity compared to the cut-based approach, the method itself is seen as a black box by some analysts, leading to distrust and increased systematic uncertainty of the final physics results.

Today, MVA is an established tool in HEP with clear advantages in analyses with dominating statistical uncertainty.

3.1.4.3 Deep Learning

The deep-learning approach is a multivariate analysis method which includes parts of the preprocessing and the feature engineering into the statistical model. It shares the same traits as the general multivariate analysis approach. Deep learning uses deep artificial neural networks and takes advantage of the large available training datasets and the massively parallel computing power provided by modern hardware.

It is long known that neural networks with one hidden layer and an arbitrary, bounded, non-constant activation function can approximate arbitrary functions in L^p[2] [18]. However, as was shown recently in [19], deep neural networks with more than one hidden layer can approximate functions of practical interest (e.g. $f(x, y) = x \cdot y$) with exponentially fewer parameters than shallow networks.

Low-level features like the measured four-momenta and calorimeter entries are directly fed to a deep artificial neural network. The deep neural network is intended to learn suitable high-level features automatically. Recent research [20] indicates that deep neural networks can extract more information from low-level features in

[2]The Lebesgue space of p-integrable functions.

comparison to high-level features created by experienced human analysts. Consequently, deep-learning approaches outperform established algorithms in many fields, as was reviewed in [21].

Physical knowledge can be incorporated into the architecture of the deep neural network. In particular, the explicit consideration of invariance properties (e.g., the invariance against rotations around the beam-line) is important.

The application of convolutional and recurrent neural networks in HEP is investigated. Adversarial networks [22] allow to selectively exclude features from the generated statistical model, to enforce a uniform signal selection efficiency in this feature [23]. This is discussed in greater detail in Sect. 3.3.2.

In the theory of representation learning [24], the improved performance is explained by the transformation of the low-level input feature space into a representation, which allows to easily solve the tasks at hand.

An illustrative example is introduced in [25]. In this work a deep neural network is presented, which can create textual descriptions of an image. The pixels of the image are transformed to an intermediate representation of the objects contained in the image by a convolutional neural network. Subsequently, the intermediate representation is transformed into a textual description of the image by a recurrent neural network.

The key concepts of representation learning are (as described in [24]):

- **distributed representations**—the number of input regions which can be distinguished grows exponentially $\mathcal{O}(2^N)$ in the numbers of parameters N;
- **depth**—the ways to re-use learned features grow exponentially with the depth of the network;
- **abstraction**—the learned features in the deeper layers are increasingly invariant to most local changes of the input;
- and **disentangling factors of variation**—the learned features represent independent properties of the input data.

On the other hand, due to the direct usage of many low-level features it is unclear how to define suitable control regions and channels. At the same time, the sensitivity to mis-modeling in Monte Carlo simulation is further increased in comparison with ordinary MVA methods, due to the reliance on a multitude of low-level features. Therefore, it is more difficult to estimate systematic uncertainties of deep neural networks in absence of calibration techniques on data. This problem can be mitigated with data-driven techniques like event-based re-weighting (see Sect. 3.3.3).

Instead of feature engineering the time is spent in architecture engineering. Finally, the field evolves quickly and there are no established and commonly accepted fields of application in HEP yet.

The usage of deep learning revolutionized many fields in recent years. Applications with a learnable distributed representation, are in particular suitable for deep-learning.

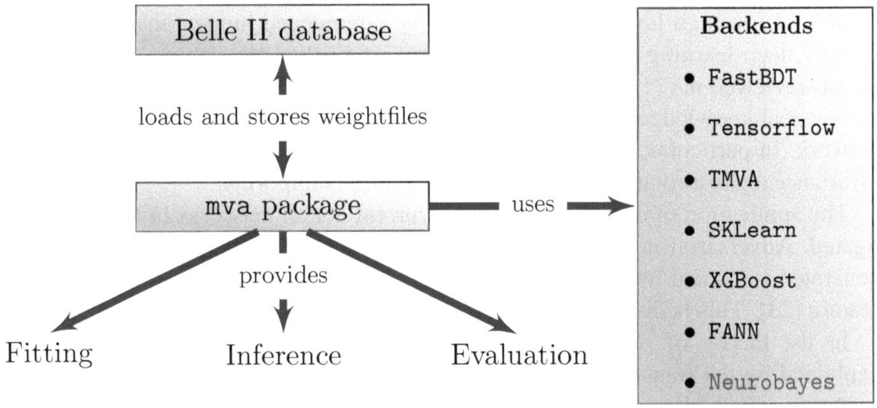

Fig. 3.3 Schematic overview of the `mva` package. It provides: an automatable fitting, inference and evaluation interface to the user; transparently loads and stores the `WeightFiles` (serialized form of the underlying statistical models) in the Belle II database; and can use the most popular MVA frameworks as backend

3.2 Implementation

The `mva` package provides a common code base for MVA related algorithms in `BASF2`. Figure 3.3 shows a schematic overview of the package. It is used in: the online reconstruction by the `FastReco` path of the trigger package; the offline reconstruction by the `tracking`, `ecl` and `klm` packages; and the mDST analysis by all MVA-based algorithms like `Continuum Suppression`, `Flavour Tagging`, and the `Full Event Interpretation`.

The `mva` package can use most of the popular MVA frameworks as backend, but it is not a wrapper[3] around existing MVA frameworks. Rather, it provides glue-code, which integrates the frameworks into `BASF2`, so that the user can focus on dealing with the framework of his choice instead of the merits of `BASF2` and `ROOT`.

For the fitting-phase, the `mva` package: converts the training data from the `ROOT`-format used by Belle II into the appropriate backend-specific format; streams the data to the backend (conserving the potential out-of-core capability of the backend); bundles the generated backend-specific `WeightFiles` (serialized form of the underlying statistical models) and all information necessary to reproduce the fitting into a Belle II `WeightFile`; and uploads the Belle II `WeightFile` into the `Belle II Conditions Database` with an optional interval of validity (a range of experiments and runs for which the `WeightFile` should be used). In addition, basic examples for all backends are provided, which the user can use as a starting-point; and advanced examples explaining state-of-the art machine

[3] A wrapper would provide a backend-agnostic interface and in consequence the lowest common denominator of all backends.

learning techniques like Bayesian hyper-parameter optimization [14]; adversarial neural networks [23]; feature importance calculation and sPlot [26, 27].

For the inference-phase, the mva package: (re-)loads the appropriate Belle II WeightFile from the Belle II Conditions Database for the current experiment and run; loads the backend-specific model; extracts the required features from the DataStore[4]; calculates the response using the model and stores the response in the DataStore.

For the evaluation, the mva package provides: plotting primitives for features (distribution, correlations); plotting primitives for the method response (distribution, ROC curves, over-fitting checks); and calculation of feature importances. The evaluation tools are backend-agnostic and can be used to compare different backends.

Finally, the package was designed with a large degree of automation in mind. The provided tools can be invoked from C++, Python and bash with similar interfaces. Fitting, inference and evaluation do not require human interaction.

3.2.1 Methods

In the following sections I briefly describe key ideas and concepts of the supported backends. However, the machine learning field evolves quickly and most of the projects will have published updated versions since the publication of this thesis. Besides the frameworks mentioned below, the mva package can be easily extended to new backends upon user request.

3.2.1.1 FastBDT

Belle II required a default multivariate classification algorithm which is: fast during fitting and application; robust enough to be trained in an automated environment; can be reliably used by non-experts; preferably generates an interpretable model and exhibits a good out-of-the box performance.

FastBDT [28] fulfills these requirements and is the current default classification method of the mva package. It implements a speed-optimized and cache-friendly implementation of the widely employed Stochastic Gradient-Boosted Decision Tree (BDT) algorithm [11], which exhibits a good out-of-the-box performance and generates an interpretable model. Furthermore, FastBDT supports: preprocessing with equal-frequency binning and purity-transformation; correct handling of missing data; and boosting to uniformity (see Sect. 3.3.2.3).

FastBDT was originally developed during this thesis to speed up the fitting and inference-phase of the Full Event Interpretation (see Chap. 4). The source-code is licensed under GPLv3 and available on github [29]. FastBDT is shipped with BASF2 and is available on all computing sites [28].

---

[4] A memory-representation of the current event.

3.2.1.2 TMVA

The TMVA library is provided by the ROOT framework. TMVA [30] is traditionally used in HEP and implements a wide range of methods: analytical methods like the Fisher discriminant; decision tree based like gradient-boosted decision trees; different forms of neural networks; probability density estimators; and rule-based approaches. It supports classification, multi-class classification and regression. Furthermore, preprocessing steps like normalization, decorrelation and Gaussianization can be performed on all or a subset of features.

TMVA is actively developed by the HEP community, and started to adopt recent developments like deep learning. Since it is shipped with the ROOT framework, it is available on all computing sites.

3.2.1.3 FANN

The Fast Artificial Neural Network (FANN) library [31] is an open-source neural network library, which implements a multi-layer neural network. Input features and the output of the network can be automatically scaled to sensible ranges. Furthermore, it supports fitting and inference on multiple CPU cores. It is shipped with BASF2 and is available on all computing sites.

3.2.1.4 NeuroBayes

NeuroBayes [32] is a closed-source, commercial implementation of a neural network. It provides an advanced feature preprocessing (equal-frequency binning, purity-transformation, b-spline fits); Bayesian regularization methods like ARD (automatic relevance determination); and evaluation capabilities including feature importance estimations. It was extensively used by the Belle experiment.

As of the time of writing it is not further developed, but legacy support is available. Due to license restrictions it is only available on some computing sites (including the KEK computing center, DESY and KIT IETP). Belle II decided to use NeuroBayes only inside belle_legacy (for example in the b2bii package) code.

3.2.1.5 Python-Based

The mva package can support all Python-based MVA frameworks. For this the package contains an API (application programming interface) defining a set of Python hook-functions, which can be used to build, fit, save, load and apply arbitrary Python-based MVA methods. By implementing the hook-functions, the user can employ all MVA frameworks which provide a Python interface.

For the most popular `Python`-based MVA frameworks, the `mva` package already predefines all the necessary hook-functions, however the user can still override them in a user-defined `Python` file.

Officially supported (i.e. examples and test-code is provided by `BASF2`) are:

- `SKLearn` [33], is part of `SciPy`,[5] it supports classification, regression, clustering, dimensionality reduction, model selection and preprocessing;
- `XGBoost` [34], is an open-source, scalable, portable and distributed Gradient Boosting library;
- `hep_ml`, an open-source machine learning library dedicated to algorithms used in HEP;
- `Tensorflow` [35], is an open-source library for (deep) neural networks using CPU and GPUs, with wide adoption by technology companies around the world;
- `Keras` [36], is an open-source, high-level interface to neural network frameworks like Tensorflow, it was developed for easy and fast prototyping of deep learning algorithms.

Other `Python`-based frameworks like `pylearn2` [37] and `theano` [38] are supported unofficially. The individual `Python`-based methods are not available on the computing sites, but can be installed by the user.

3.2.2 Benchmark: $D^0 \rightarrow K^-\pi^+\pi^0$

Separating correctly reconstructed particles from background is a frequent classification task during **mDST and n-tuple analysis**. The reconstruction of the decay $\mathrm{D}^0 \rightarrow \mathrm{K}^-\pi^+\pi^0[\rightarrow \gamma\gamma]$ is used as a benchmark classification in the sections below. This decay contains the most abundant final-state particles in the detector K, π and γ, and has a rich phase-space structure with various intermediate resonances like ρ and K^*. Furthermore, with a branching fraction of $(14.3 \pm 0.8)\%$ [39], it is also the most common decay of the D^0.

The decay is reconstructed from a K^- candidates with a kaon probability above 0.5, a π^+ candidate with a pion probability above 0.5, and a $\pi^0 \rightarrow \gamma\gamma$ candidate with χ^2 probability of a mass-constrained vertex fit above 0.1. Further selection criteria are applied on the γ candidates. The invariant mass of the final D^0 meson has to be between 1.7 and 1.9 GeV. A vertex fit is performed, and the χ^2 probability of the fit has to be above 0.1. The following input features are used if not stated otherwise: the invariant mass of the D^0 meson; the pairwise invariant masses of the decay products $\mathrm{M}_{\mathrm{K}^-\pi^+}$, $\mathrm{M}_{\mathrm{K}^-\pi^0}$, and $\mathrm{M}_{\pi^+\pi^0}$; particle identification information of K^- and π^+; D^0, K^-, π^+ and π^0 momenta; D^0, K^- and π^+ vertex information; and the χ^2 fit probability of D^0 (vertex-fit), K^+ (track-fit), π^+ (track-fit) and π^0 (mass-vertex-fit).

[5] A Python-based ecosystem, of open-source software for mathematics, science, and engineering.

In total 745,000 charged B^+B^- Monte Carlo simulated events were processed, and split into a training (355,000 events) and an independent validation sample (390,000 events). The signal fraction in both samples is 6.6%.

3.2.2.1 Comparison

The benchmark introduced in Sect. 3.2.2 was used to compare the performance (in terms of fitting runtime, inference runtime, classification quality and `WeightFile` size) of different backends.

Two sets of classifiers were compared among each other on an Intel(R) Core(TM) i7-4770 CPU (@ 3.40 GHz) with a main memory of 32 GB. To ensure a sound comparison, all methods were restricted to one core and typical hyper-parameters were chosen.

- Stochastic gradient-boosted decision tree implementations (`FastBDT`, `TMVA`-BDT, `SKLearn`-BDT, `XGBoost`) with the same hyper-parameters (one hundred trees with a depth of three).
- Artificial neural network implementations (`FANN`, `TMVA`-NN, `SKLearn`-NN, `Tensorflow`, `NeuroBayes`) with the same hyper-parameters (one hidden layer with 29 neurons and maximally one hundred iterations through the dataset during the fitting). The input features were normalized (shifted to a mean of zero, and scaled to a standard deviation of one).

The presented results depend on the chosen preprocessing steps, hyper-parameters and the specific task. In consequence, this comparison is only a rough guideline of the strengths and weaknesses of the backends. The results are summarized in Table 3.1.

The BDT methods are faster than the Neural Networks during the fitting-phase, at the same time they outperform the Neural Networks in the quality of the separation (measured by the area under the receiver operating characteristic AUC ROC [40]). `FastBDT` is the fastest method and obtains the best AUC ROC score on the benchmark. The convergence of `FANN` for this benchmark was unstable i.e. the obtained ROC values fluctuated strongly.

The runtime required to load the validation data from disk dominates the inference-phase (as can be seen from the `Trivial` method, which only loads the validation data and does not perform computations). To ensure a significant result the validation dataset statistics was increased by factor 10. Contrary to the naive expectation that the Neural Networks are faster during the inference because they rely only on linear algebra operations, which can be performed very efficiently on modern hardware; `FastBDT` is the fastest method. `SKLearn`-BDT is nearly as fast as `FANN` and `SKLearn`-NN, faster than `TMVA`-NN and `Tensorflow`. `NeuroBayes` is the slowest contestant.

Several effects have to be considered to explain the unexpected result: preprocessing steps can be very expensive (e.g. `NeuroBayes` does several non-linear transformations as preprocessing); careful optimizations for caching and pipelining have a large impact on modern hardware (as can be seen by the results of `FastBDT`).

Table 3.1 Comparison of the different backends using the benchmark $D^0 \rightarrow K^-\pi^+\pi^0$. The first column shows the fitting time on 28 features and 355,000 events measured in seconds. The second column shows the inference time on 28 features and 3,900,000 events measured in seconds. The third column shows the area under the receiver operating characteristics as a measure of the classification quality (more is better). Different quantities can be used to construct the receiver operating characteristic, here the signal efficiency and purity are used. The last column shows the size of the `WeightFile` in kilobytes. All measurements were performed 10 times, the minimal achieved time in fitting and inference, and the average AUC ROC is stated in the table. The `Trivial` method does nothing during the fitting-phase (time equals overhead of the `mva` package), and returns the signal-fraction during the inference-phase (time equals overhead due to data loading)

Method	Fitting time in s	Inference time in s	AUC ROC	`WeightFile` size in KB
`Trivial`	0.2	4.9	0.066	2
Stochastic gradient boosted decision tree				
`FastBDT`	3.7	6.9	0.435	58
`SKLearn-BDT`	32.1	7.8	0.429	69
`XGBoost`	18.0	11.4	0.415	34
`TMVA-BDT`	19.8	16.5	0.297	101
Artificial neural network				
`SKLearn-NN`	27.6	7.2	0.401	32
`Tensorflow`	201.9	9.4	0.399	30
`NeuroBayes`	112.3	75.4	0.377	182
`FANN`	50.6	7.1	0.316 ± 0.061	21
`TMVA-NN`	510.6	16.8	0.156	53

In consequence, even the same algorithm can have very different runtimes depending on the implementation (as can be seen by the different BDT implementations, which all implement stochastic gradient-boosting).

3.2.3 Conditions Database Integration

The `Belle II Conditions Database` contains experiment and run-dependent information required for the processing of Monte Carlo simulated events and detector-data. Traditionally machine parameters like: beam energies; luminosity; detector alignment and status; and calibration constants are stored in the conditions database. Due to the expected extensive usage of MVA methods in Belle II, and their dependence on the run-dependent background conditions, the `WeightFiles` of MVA methods used in the online and offline reconstruction (likely also mDST analysis) can be fitted per experiment (or run) and stored in the `Belle II Conditions Database`.

Furthermore, the database provides a revision system for `WeightFiles`, an infrastructure to automatically distribute the required `WeightFiles` to the computing sites, and ensures reproducibility.

The `mva` package transparently loads and stores `WeightFiles` in the `Belle II Conditions Database`. Optionally, the `WeightFiles` can be loaded from and stored on disk in the `ROOT` and `XML` format. Depending on the backend, used method and its hyper-parameters, the `WeightFiles` usually require between 10 KB and 1 MB disk-space. Table 3.1 states the size of the `WeightFiles` for the different backends in the `ROOT` format.

3.2.4 *Automatic Evaluation*

Evaluating the fitted models is important: to spot possible issues in the input features; to check for convergence of the algorithm; to avoid biases due to over-fitting and under-fitting; and compare different models to one another. The automatic evaluation of the `mva` package creates a summary in form of a `PDF` document, which can later be examined by the user. It is also possible to observe the quality of an MVA method in an automated fitting environment and to sent emails to quality control if the quality drops below a predefined threshold.

The `PDF` summary of the `mva` package contains: the configuration of the used methods; the correlation-matrix of all features for signal and background; the importance of each feature; the distribution of each feature and spectator; the receiver operating characteristics on the training and an independent test datasets for all methods; the distributions on the training and an independent test dataset of the response for all methods; and the diagonal plots, which show the purity of the binned response for all methods. Additional plots (like `tSNE` plots, and control regions plots for different response-cuts) are provided on demand.

3.3 Applications

This section summarizes the usages of machine learning and the `mva` package in the Belle II collaboration. Section 3.3.1 describes applications during mDST analysis; these were developed during this thesis and many are used in the analysis $\mathrm{B} \to \tau\nu_\tau$. Furthermore, the `mva` package is used extensively during online and offline reconstruction: in the identification of K_L^0 particles from clusters in the KLM; the energy and shape determination of clusters in the ECL; and the track finding and fitting algorithms in the CDC and VXD.

This thesis focused in particular on the development of basic building blocks for complex algorithms. Often, some features have to be reserved for subsequent steps in the analysis like the extraction of the signal yield using a fit. Therefor, discriminators with an enforced uniform signal selection efficiency are investigated in

Sect. 3.3.2. Section 3.3.3 introduces data-driven techniques, which are a key to reduce the dependency on Monte Carlo simulation. Finally, Sect. 3.3.4 outlines algorithms used to optimize the hyper-parameters of an MVA method.

3.3.1 Analysis Algorithms

All multivariate algorithms used in the `analysis` package of `BASF2` are based on the `mva` package. In this section I briefly describe the three most important algorithms, which were implemented during this thesis (Sect. 3.3.1.1) or during supervised master's theses (Sects. 3.3.1.2 and 3.3.1.3).

3.3.1.1 Full Event Interpretation

The Full Event Interpretation (`FEI`) is used for hadronic and semileptonic tagging. It automatically reconstructs hadronic and semileptonic B meson decay-chains and infers a signal probability for each reconstructed candidate. This algorithm is used in the measurement of branching fraction of rare decays with missing kinematic information like $B \rightarrow \tau\nu_\tau$, $B \rightarrow \ell\gamma\nu$ and $B \rightarrow \nu\nu$.

The signal probability is calculated using a hierarchical network of multivariate classifiers, which is visualized in Fig. 4.3. A boosted decision tree (BDT) based on `FastBDT` (see Sect. 3.2.1.1) is trained for each final state particle and each decay-channel of the intermediate particles. For each particle candidate a signal probability is calculated using the corresponding BDT. The signal probabilities of the daughter particles are used as features in the training of the BDT of a composite intermediate particle. In this manner the complete information about the entire decay-chain is encoded in the signal probability of the final B candidates.

The training, application and evaluation of the more than 100 BDTs employed by the `FEI` is fully automatized with the tools provided by the `mva` package.

The FEI is described in greater detail in Chap. 4.

3.3.1.2 Flavour Tagging

`BASF2` includes two flavour tagging algorithms, which distinguish the decay products of a B^0 meson from a $\bar{B}^0$ meson. Flavour-tagging is used in time dependent CP violation analyses like the measurement of $\mathcal{A}_{CP}(B^0 \rightarrow J/\psi K^0_s)$.

The category-based flavour tagger trains either boosted decision trees (see Sect. 3.2.1.1) or neural networks (see Sect. 3.2.1.3) to identify certain flavour-specific signatures (so-called categories) like primary lepton decays or $b \rightarrow c \rightarrow s$ decay-chains [41]. Each classifier receives hand-crafted features, which are known to be correlated to the flavour-specific signature. The output of the category classifiers are combined by another BDT or NN to the final flavour tagging information (1 for

B^0 and -1 for $\bar{B}^0$) and the estimated uncertainty on the tag (0 for a random choice and 1 for absolute certainty).

The deep learning-based (see Sect. 3.1.4.3) flavour tagger receives only low-level features as its input [42]. The four momenta, POCA (point of closes approach of the track to the beam pipe) and PID (particle identification) information of all tracks associated to the decay products. The quantities are fed into a deep neural network ordered by the total momentum of the associated track and separated by charge. It is assumed, that the deep neural network learns the relevant flavour-specific signatures and correlated high-level features automatically.

The training and application of both taggers is automatized with the tools provided by the `mva` package.

The performance of the flavour tagger based on deep learning is out-of-the-box equal to the traditionally used category-based flavour tagger. Both taggers can be evaluated and calibrated on data using the `b2bii` package.

3.3.1.3 Continuum Suppression

Most analyses at the Belle II experiment investigate the decay of B mesons from an $e^-e^+ \rightarrow \Upsilon(4S) \rightarrow B\bar{B}$ decay. An important background to these analyses are continuum events $e^-e^+ \rightarrow q\bar{q}, \tau^-\tau^+, \mu^-\mu^+, e^-e^+$.

Historically, the difference in the event topology was used to suppress continuum events. Due to the low momentum of the B mesons in the center of mass (CMS) frame, the topology of a $B\bar{B}$ event is spherical in the CMS frame. The continuum events show a twofold cone-shaped back-to-back structure in the CMS frame. Hand-crafted high-level features were designed to systematically describe the event topology like the CLEO cones [43] and the KSFW (Kakuno Super Fox Wolfram) moments (see [44] and [45, Chap. 9]). Figure 3.4 shows example event topologies of a continuum event in Fig. 3.4a, a possible signal-decay $B^- \rightarrow \tau^-\bar{\nu}_\tau$ event in Fig. 3.4b and a B^+B^- event in Fig. 3.4c. As can be seen from the examples, the signal event topology is more similar to the continuum than the $B\bar{B}$ event, due to the one-prong decay on the signal-side. Hence, it is important to consider the specific signal-side in the continuum suppression algorithm.

`BASF2` includes two continuum suppression algorithms.

The traditional continuum suppression algorithm uses a boosted decision tree[6] (see Sect. 3.2.1.1), which was fitted to suppress the continuum background, using hand-crafted high-level features (see Sect. 3.1.4.2).

The deep-learning-based (see Sect. 3.1.4.3) continuum suppression algorithm receives in addition low-level features as its input [46]. The four-momenta and POCA of the charged particles are transformed into the CMS frame and rotated with respect to the thrust axis. They are fed to a deep neural network in spherical coordinates ordered by the magnitude of the momentum of the particles and separated by the charge and their assumed affiliation to the signal or tag side.

---

[6]FastBDT: One hundred trees with a depth of three.

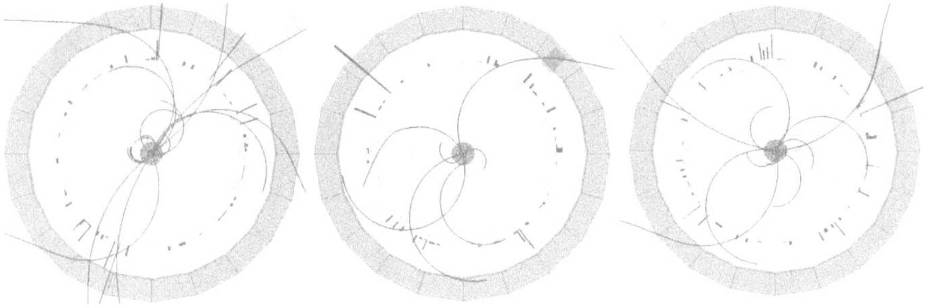

(a) $e^-e^+ \rightarrow c\bar{c}$ (b) $B^- \rightarrow \tau^-\bar{\nu}_\tau$ (c) $e^-e^+ \rightarrow \Upsilon(4S) \rightarrow B^+B^-$

Fig. 3.4 Event displays showing the event topologies of different event types. The difference of the spherical topology of $\Upsilon(4S)$ and the twofold cone-shaped back-to-back structure of continuum events is exploited by hand-crafted continuum suppression features to separate them from one another

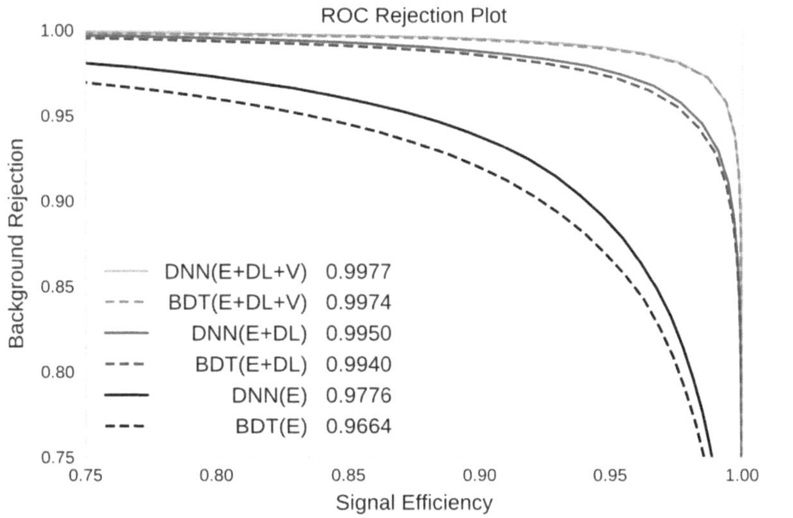

Fig. 3.5 Comparison between different feature sets: traditional hand-crafted features E (30), low-level kinematic features DL (440) and low-level vertex features (60). The addition of low-level features increases the quality of the separation. Taken from [46]

Both continuum suppression algorithms are trained using a dataset containing signal events (instead of generic $B\bar{B}$ events) and continuum events. The addition of low-level features leads to a significant improvement in the separation (see Fig. 3.5). No significant difference between a BDT and a simple deep neural network were observed. However, more advanced network architectures lead to further improvements [46].

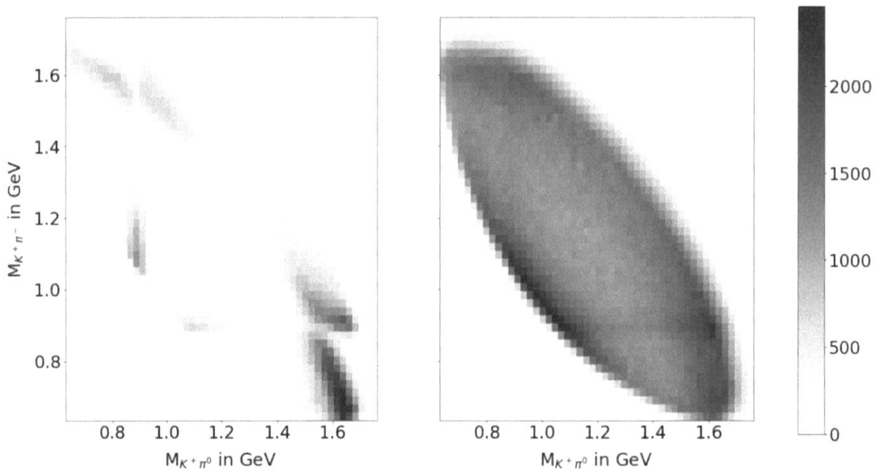

Fig. 3.6 Dalitz plane of the benchmark decay $D^0 \rightarrow K^-\pi^+\pi^0$. (Left) correctly reconstructed D^0; showing the rich substructure due to the intermediate resonances in the decay. (Right) wrongly reconstructed D^0

3.3.2 *Uniformity-Constraint*

Physics analyses often employ MVA classifiers to improve the signal-to-noise ratio using multiple *features*, which can separate between signal and background. On the other hand, many analyses extract physical observables by fitting a signal and background model to one or more *fit-variables*. A typical example is a two-dimensional fit to the Dalitz plane of the benchmark decay $D^0 \rightarrow K^-\pi^+\pi^0$ to extract the fraction, mass and width of intermediate resonances in the decay. In this example, the *fit-variables* are the invariant masses of the $K^-\pi^+$ system $M_{K^-\pi^+}$ and the $K^-\pi^0$ system $M_{K^-\pi^0}$; whereas the remaining quantities described in Sect. 3.2.2 are used as *features* in the classifier. Figure 3.6 shows the signal and background distribution of the benchmark decay in the *fit-variables*.

It is advantageous to require a uniform selection efficiency of the classifier for signal and background in the *fit-variables*, separately.

- A non-uniform selection efficiency of background events creates an artificial peaking background if a cut on the classifier output is performed. The shape of the background model depends on the cut on the classifier output. This can lead to large systematic uncertainties if the background model cannot be checked independently in a control region.
- A non-uniform selection efficiency of signal events changes the shape of the signal model depending on the cut on the classifier output. This increases the reliance on Monte Carlo simulation, because usually the signal model cannot be checked independently in a control region.

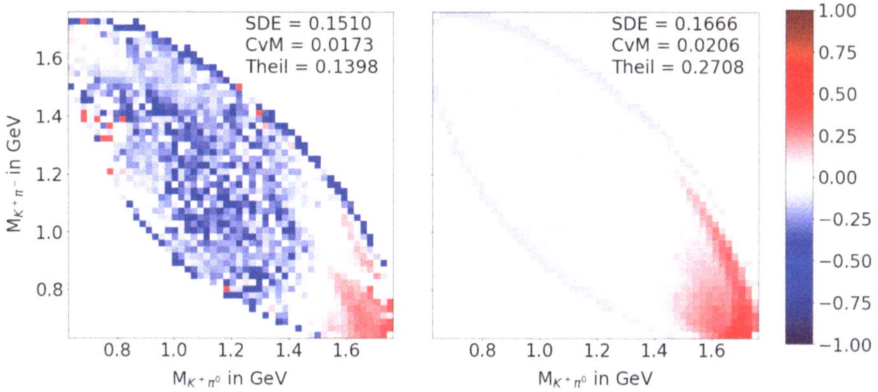

Fig. 3.7 **Baseline BDT**: Deviation from the average selection efficiency for the cut on the BDT response, which maximizes the signal-to-noise ratio. The left (right) plot shows signal (background) D^0 candidates

- Contrastingly, a non-uniform selection efficiency of signal and background combined is desired, because otherwise the signal-to-noise ratio would not increase by cutting on the classifier output.

In the following I describe three algorithms, which construct classifiers with a uniform selection efficiency in the *fit-variables*. They are compared to two baseline models with non-uniform selection efficiency (Sect. 3.3.2.1) on the benchmark (Sect. 3.2.2).

The performance is compared in terms of the classification quality, measured as usual by the area under the receiver operating characteristic curve[7] (AUC ROC), and the uniformity of the selection efficiency. As described in [47], the uniformity of the selection efficiency can be measured by different approaches using: the standard deviation of the efficiency in bins (SDE); the Theil index (Theil); and the Cramér-von-Mises similarity (CvM). For all measures, a lower value indicates a more uniform selection efficiency.

All investigated algorithms were included in the `mva` package during this thesis.

3.3.2.1 Baseline

Two baseline models trained on the benchmark are used: a boosted decision tree implemented in `hep_ml` (BDT) and a neural network implemented in `tensorflow` (NN). The uniformity of the selection efficiency of the BDT and the NN is visualized in Figs. 3.7 and 3.8, respectively. As can be seen from the figures, both algorithms have a non-uniform selection efficiency for both signal and background.

---

[7]Different quantities can be used to construct the receiver operating characteristic, here the signal efficiency and background rejection are used.

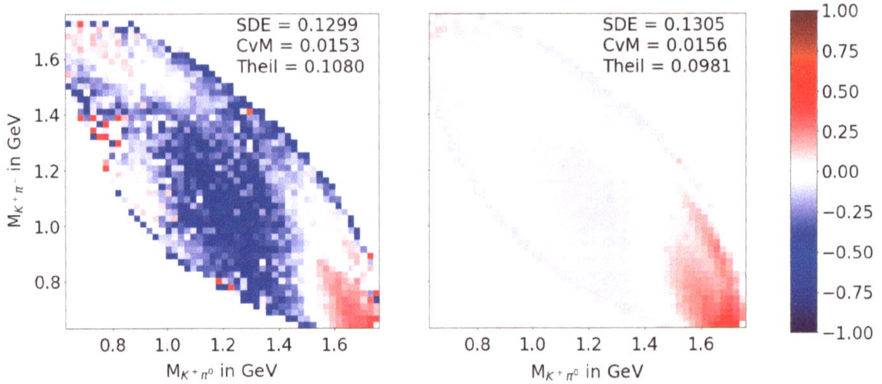

Fig. 3.8 **Baseline NN**: Deviation from the average selection efficiency for the cut on the neural network response, which maximizes the signal-to-noise ratio. The left (right) plot shows signal (background) D^0 candidates

3.3.2.2 Feature Drop

The simplest method to enforce a uniform selection efficiency is to remove all *features* which are dependent on the *fit-variables*. The dependency can be quantified by: **linear correlation coefficients** (linear and ignores multivariate correlations); **statistical hypothesis tests** for two features being independent [48] (non-linear and ignores multivariate correlations); or the feature importance of the *features* in a **multivariate regression** to predict the *fit-variable* (non-linear and considers multivariate correlations).

Using handcrafted features designed to be uncorrelated to the *fit-variables*, e.g., the beam-constrained mass and ΔE as a parametrization of the four-momentum of a B meson, can further improve this approach.

A major drawback of this method is that the *features* with high separation power are often also dependent on the *fit-variables*. Therefore, one potentially looses a lot in terms of classification quality as is evident from Figs. 3.13 and 3.14. Furthermore, as can be seen from Figs. 3.9 and 3.10 even aggressively removing dependent *features* does not guarantee a uniform selection efficiency. Empirically, there is always some dependency between the *features* and the *fit-variables*, so enforcing an exact uniform selection efficiency would require removing all *features* from the training.

To sum up, more powerful techniques are required to enforce a uniform selection efficiency.

3.3.2.3 Boosting to Uniformity

As described in [47], boosting (re-weighting of events depending on the output of a classifier) can be used to enforce a uniform selection efficiency. This technique

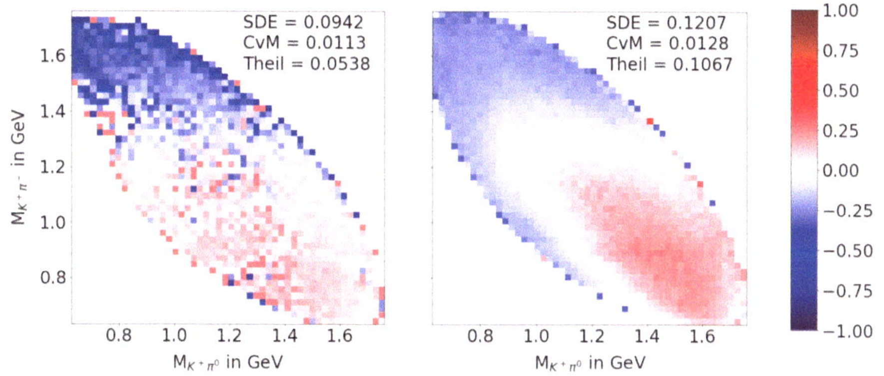

Fig. 3.9 BDT with FeatureDrop: Deviation from the average selection efficiency for the cut on the BDT response, which maximizes the signal-to-noise ratio. The left (right) plot shows signal (background) D^0 candidates. The BDT was trained on a subset of *features* without kinematic information of the daughter particles

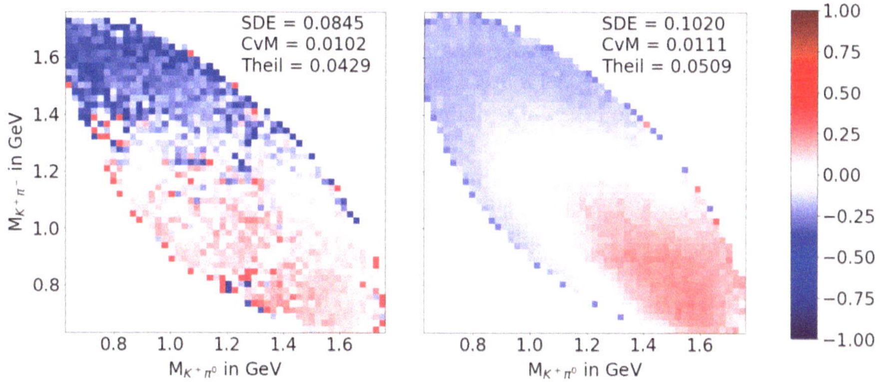

Fig. 3.10 NN with FeatureDrop: Deviation from the average selection efficiency for the cut on the NN response, which maximizes the signal-to-noise ratio. The left (right) plot shows signal (background) D^0 candidates. The NN was trained on a subset of *features* without kinematic information of the daughter particles

can be naturally incorporated in a boosting algorithm, like it is used for stochastic gradient-boosted decision trees. The loss-function is extended by an additional term $\mathcal{L}_{\text{flat}}$, which penalizes a non-uniform selection efficiency

$$\mathcal{L}_{\text{flat}} = \alpha \, \text{CvM}_S + \beta \, \text{CvM}_B, \tag{3.4}$$

where CvM is the Cramér-von-Mises similarity introduced in Sect. 3.3.2, and α and β are new hyper-parameters controlling the strictness of the uniformity constraint.

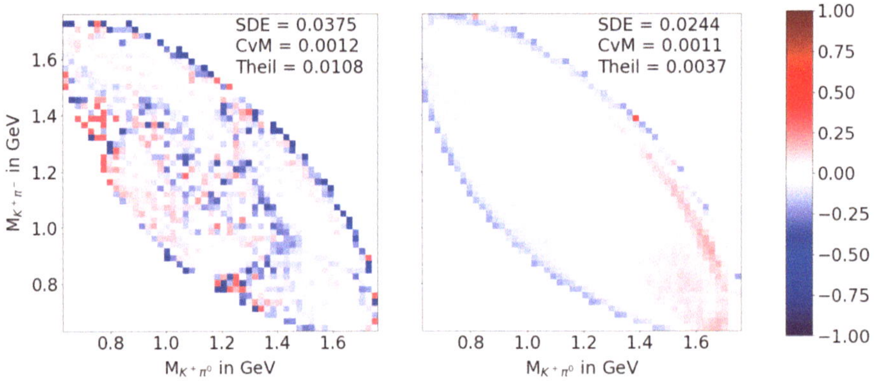

Fig. 3.11 BDT with uniformity boosting: Deviation from the average selection efficiency for the cut on the BDT response, which maximizes the signal-to-noise ratio. The left (right) plot shows signal (background) D^0 candidates

The boosting algorithm will increase (decrease) the weight of events in bins[8] with lower (higher) selection efficiency compared to the inclusive distribution.

As can be seen from Fig. 3.11 this algorithm yields a nearly uniform selection efficiency on the benchmark while maintaining a high classification quality (see Fig. 3.13).

This algorithm was implemented into `FastBDT` (see Sect. 3.2.1.1) during this thesis, and is also (independently) available in the `hep_ml` package.

3.3.2.4 Adversarial Neural Networks

Adversarial neural networks are commonly used in generative neural networks [22]. Louppe et al. [23] applied the technique to create a neural network classifier $f(\vec{x}, \vec{w})$, whose output s is independent of the values $\vec{z}$ of some nuisance parameters Z

$$p(f(\vec{x}, \vec{w}) = s|\vec{z}) = p(f(\vec{x}, \vec{w}) = s|\vec{z}^\star) \tag{3.5}$$

for all $\vec{z}, \vec{z}^\star \in Z$ and all possible outputs s of the classifier. Louppe et al. [23] proofs that the obtained classifier is both optimal[9] and fulfills Eq. 3.5, if such a solution exists. Equation 3.5 implies that $f(X)$, Z are independent random variables, therefore a cut on the output of f has a uniform selection efficiency in Z by construction.

The technique was adapted during this thesis to create a uniform discriminator by identifying the nuisance parameters with the *fit-variables*. As explained in Sect. 3.3.2, a uniform selection efficiency is desired for signal and background separately. This can be achieved by additionally conditioning on the event class c

[8] A joint region in the *fit-variables*.

[9] In the sense defined in [23].

$$p(f(\vec{x}, \vec{w}) = s|\vec{z}, c) = p(f(\vec{x}, \vec{w}) = s|\vec{z}^\star, c) \tag{3.6}$$

for all $\vec{z}, \vec{z}^\star \in Z, c \in \{S, B\}$ and all possible outputs s of the classifier. In particular the random variables $f(X)$, Z are now conditionally independent instead of independent.

For each pair (z_i, c_i) of *fit-variable* z_i and class c_i, an adversarial neural network is trained in parallel to the neural network f responsible for the classification. Each adversarial neural network learns to predict the probability density function $p(z_i|s, c)$ of its *fit-variable* and class; using the output s of f as the sole input. This can be accomplished by minimizing the loss-function $\mathcal{L}_i = -\log(p(z_i|s, c_i))$. The sum of all adversarial loss-functions $\mathcal{L}_i$ is subtracted from the loss-function $\mathcal{L}_f$ of the classification neural network

$$\mathcal{L} = \mathcal{L}_f - \alpha \sum_i \mathcal{L}_i, \tag{3.7}$$

where α is a new hyper-parameter controlling the strictness of the uniformity constraint. Hence, the adversarial neural networks learn to extract any information on the *fit-variables* contained in the output of f. At the same time the classification neural network is penalized if the adversarial neural networks succeed in doing so. All networks are trained simultaneously until convergence.

As can be seen from Fig. 3.12 this algorithm yields a nearly uniform selection efficiency on the benchmark while maintaining a high classification quality (see Fig. 3.14).

The usage of adversarial neural networks as presented in this section is key to a successful deployment of deep learning algorithms in HEP, because it can prevent the network from learning specific *fit-variables*. Examples include: The decay length difference Δz in CP measurements, and the beam-constrained mass $M_{\rm bc}$ in hadronic tagging algorithms.

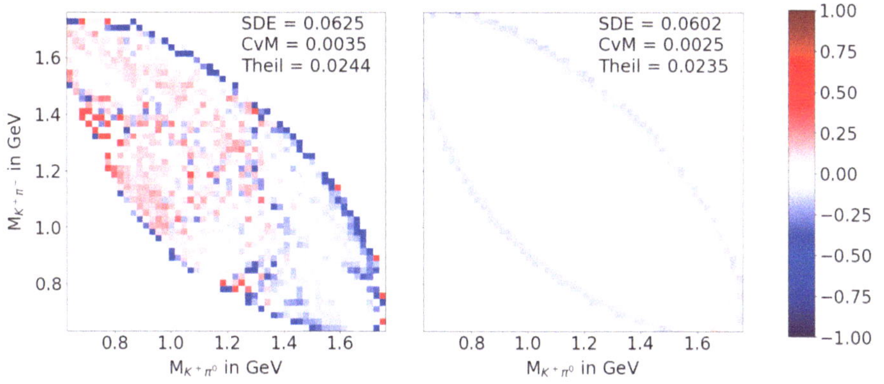

Fig. 3.12 **NN with adversary**: Deviation from the average selection efficiency for the cut on the NN response, which maximizes the signal-to-noise ratio. The left (right) plot shows signal (background) D^0 candidates

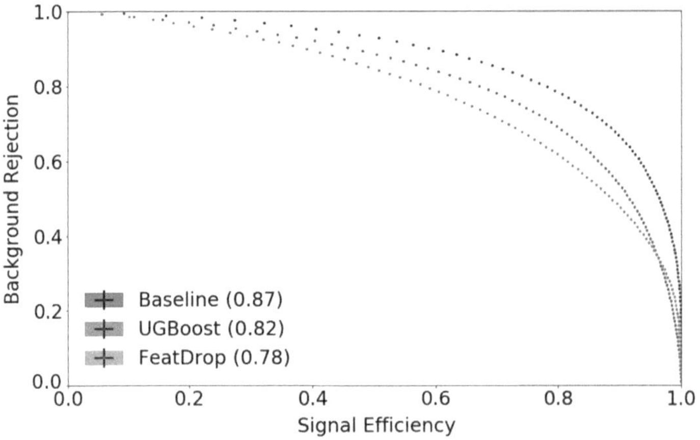

Fig. 3.13 Receiver operating characteristic (ROC) of: an ordinary BDT (baseline), a BDT using uniform boosting (UGBoost), an ordinary BDT trained on a subset of *features* without kinematic information of the daughter particles (FeatDrop). The number in parenthesis states the area under the corresponding ROC curve

3.3.2.5 Conclusion

It is possible to ensure a uniform selection efficiency for signal and background separately: For BDTs the boosting to uniformity technique, and for NNs the adversary technique can be successfully employed. Both yield similar results in terms of classification quality and uniformity. The traditional Feature Drop approach is inferior.

The classification quality of the different algorithms, is compared in Fig. 3.13 for the BDT and in Fig. 3.14 for the NN. The uniformity constraint diminishes the classification quality, because information dependent on the *fit-variables* cannot be fully exploited by the classifiers.

Further algorithms exist to ensure a uniform selection efficiency. Dolen et al. [49] used hand-crafted features to design decorrelated taggers for jet classification, and investigated the possibility to automatize the process with principal component analysis. However, the algorithm (as presented in [49]) still requires human input and is only shown to work with four *features* and one *fit-variable*. Stevens and Williams [50] introduced the first boosting to uniformity technique for boosted decision trees (distinct from the one presented in this thesis), and previously [27] presented already an approach based on event re-weighting by building a tree of neural networks. Both approaches suffer from a significant runtime performance drawback, because many classifiers have to be trained.

Finally, uniformity-constrained classifiers can be used to implicitly prevent using information available only on MC. Here, the *fit-variable* is a binary variable containing 0 for MC and 1 for detector data. For instance, the adversarial neural network is trained to distinguish MC and detector data, and the neural network responsible

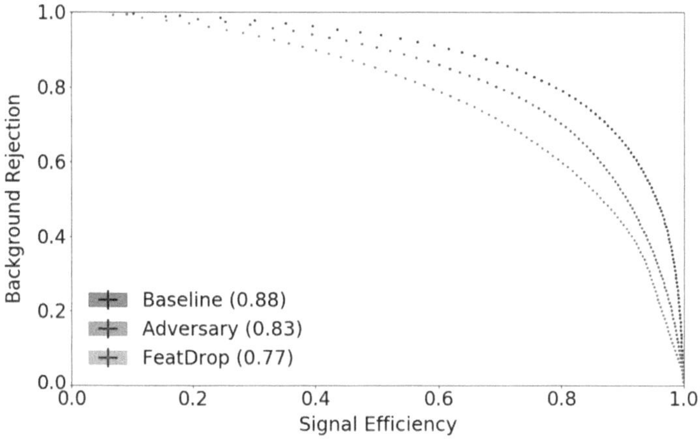

Fig. 3.14 Receiver operating characteristic (ROC) of: an ordinary NN (baseline), a NN with adversary (Adversary), an ordinary NN trained on a subset of *features* without kinematic information of the daughter particles (FeatDrop). The number in parenthesis states the area under the corresponding ROC curve

for the classification is punished if this is possible. The additional conditioning on the event class c is usually not desired (and usually not possible on data), because signal MC is easily distinguishable from detector data, if the signal is rare. The next section describes further data-driven techniques.

3.3.3 Data-Driven

Many analyses in HEP employ Monte Carlo (MC) simulation to optimize cuts; determine the shape of signal and background components; and train multivariate methods. The simulated *MC events* are usually carefully calibrated and correction factors for the particle identification performance and branching fractions are used to ensure a sound modeling of the *detector data*. In addition, control regions and channels can be used to cross-check the simulation.

As can be seen from Fig. 3.15, already the well-known benchmark decay (Sect. 3.2.2) exhibits large differences between *detector data* and the *Monte Carlo* expectation. The benchmark decay was reconstructed on *Monte Carlo* and *detector data* using the `b2bii` package. The main differences arise from the poor description of the continuum component, that is non-resonant processes like $e^+e^- \to c\bar{c}$ (see Sect. 2.3.3.3).

Data-driven techniques can be used to reduce the reliance of analyses on *Monte Carlo* simulation. Hereby, the desired information is determined directly from data by exploiting physics knowledge.

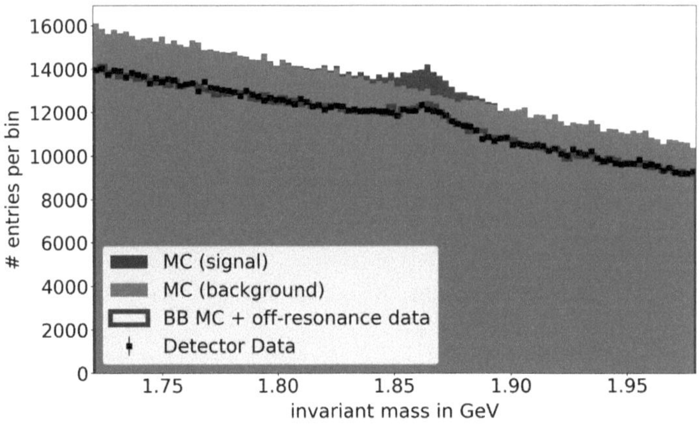

Fig. 3.15 The invariant mass of candidates reconstructed from the benchmark decay $D^0 \rightarrow K^-\pi^+\pi^0$. The filled histograms show the expected distribution from *Monte Carlo* simulation for signal (dark gray) and background (light gray). The black points show the distribution obtained from *detector data*. The overall normalization differs from the *MC expectation*. There are two main background sources: $\Upsilon(4S)$ events, which are usually well understood; and continuum events, which have large uncertainties in the description of the hadronization and the branching fractions of the subsequent decay-chains. The gray line shows *Monte Carlo* expectation, where the continuum component was replaced by off-resonance detector data. This distribution is in agreement with the *detector data*, indicating that the main difference is due to the continuum description

In the remainder of this section I introduce several data-driven techniques which aim to reduce the dependency on *Monte Carlo* simulation during the fitting-phase of multivariate methods. The first three techniques are originally described in [27]. All of them were investigated and implemented in the `mva` package during this thesis and a supervised bachelor's thesis [51].

3.3.3.1 Baseline

As a baseline for the following studies, a boosted decision tree (BDT) was trained to distinguish signal and background candidates on the benchmark decay. Figure 3.16 shows the distribution of the invariant mass of the benchmark decay after a cut on the response of the BDT. The cut was chosen to optimize the signal-to-noise ratio on *Monte Carlo*. The observed signal-to-noise ratio on data is 31.78. It was extracted by fitting a background-only model to the observed distribution on *detector data* excluding the signal peak between 1.8 and 1.9 GeV.

The training of the same boosted decision tree is repeated in the following sections using different data-driven techniques. Throughout the text the applied cuts on the response of the BDTs are always chosen to maximize the signal-to-noise ratio on *MC events* and the cited signal-to-noise ratio is always extracted on *detector data* as described above.

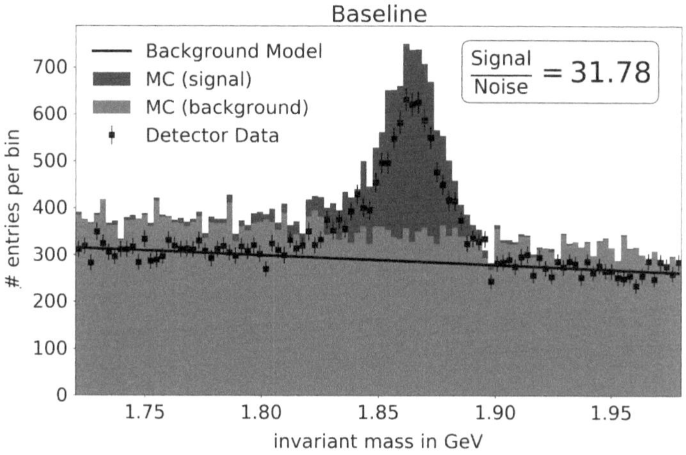

Fig. 3.16 The invariant mass distribution of the benchmark decay $D^0 \rightarrow K^-\pi^+\pi^0$, after a cut on the response of a boosted decision tree was applied. The BDT was trained on *MC events*. The filled histograms show the expected distribution from *Monte Carlo* simulation for signal (dark gray) and background (light gray). The black points show the distribution obtained from *detector data*. The black line shows the background model fitted to the *detector data*

To ensure a sound comparison, only features independent of the invariant mass were included in the training, otherwise a possible peaking background contribution could distort the evaluation on *detector data*. This corresponds to the traditional Feature Drop ansatz explained in the last section.

3.3.3.2 Event Re-weighting

The main idea of **event re-weighting** is to encode the difference of *MC events* and *detector data* in the response of a multivariate classifier. A classifier is trained to distinguish *MC events* and *detector data*. Its response is the probability $p = \frac{\text{detector}}{\text{detector+MC}}$ of selecting a *detector event* from the whole sample. An *MC event* with a high (low) probability is likely (unlikely) to appear in *detector data*, hence we want to increase (decrease) the importance of this event. The importance of each *MC event* is defined by a weight $w = \frac{\text{detector}}{\text{MC}} = \frac{p}{1-p}$ (see [27]).

The larger the differences between *MC events* and *detector data*, the better the classification quality of the classifier. In the absence of mis-modeling, the classifier would not be able to distinguish *MC events* and *detector data*.

In contrast to simple scaling factors derived from a binned distribution, this method takes multivariate dependencies, between all features used in the classifier, into account.

Subsequently, the user can train another multivariate classifier to distinguish signal from background on the re-weighted MC dataset. In particular deep learning methods, which typically employ numerous low-level features can profit from this technique

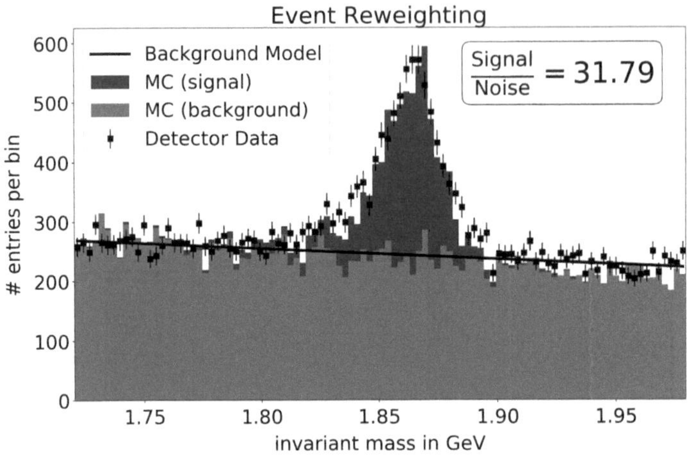

Fig. 3.17 The invariant mass distribution of the benchmark decay $D^0 \to K^-\pi^+\pi^0$, after a cut on the response of a boosted decision tree was applied. The BDT was trained on re-weighted *MC events*. The filled histograms show the expected distribution from *Monte Carlo* simulation for signal (dark gray) and background (light gray). The black points show the distribution obtained from *detector data*. The black line shows the background model fitted to the *detector data*

(see Sect. 3.1.4.3). Furthermore, the derived weights w can be used for all tasks performed on MC like determining cuts and creating plots.

This methods cannot compensate differences in phase-space regions where no *MC events* exist. Also, the method is not aware of signal and background, and could re-weight signal (background) events from MC to match background (signal) events in data.

Alternatively, one can use a uniformity-constraint classifier to mitigate differences between *MC events* and *detector data* as described in Sect. 3.3.2.5. This approach changes the underlying classifier itself, hence it cannot be automatized for arbitrary classifiers.

The **event re-weighting** technique was implemented in the `mva` package, and is automatically applied as soon as the user provides the necessary *detector data*. Figure 3.17 shows the invariant mass distribution of the benchmark decay, after a cut on a BDT trained on re-weighted *MC events*. The achieved signal-to-noise ratio does not differ significantly from the baseline (see Fig. 3.16). The re-weighted expectation from MC is in agreement with the measured spectrum on detector data.

3.3.3.3 Sideband-Subtraction

Training a multivariate classifier directly on data is challenging, because usually there are no pure signal and background samples available. However, in some situations signal-enriched and signal-free phase-space regions can be identified. These regions can be used to train a multivariate classifier on data, by statistically subtracting the background in the signal-enriched region using events from a signal-free region with a negative weight.

Sideband-Subtraction requires the definition of three phase-space regions: a **signal-enriched region**, which contains signal and background events; a **background-region**, which contains only background events; and a **negative signal-region**; which contains only background events indistinguishable[10] from the background events in the signal-enriched region. In addition, the method requires the knowledge of the expected number of signal events in the signal-enriched region e.g. from *Monte Carlo*.

Since the ground truth is not known for individual *detector events*, all events (including the background events) in the signal-enriched region are used as signal in the training. To counteract the background events contained in the signal region, the background events from the negative-signal region are used as signal as well, but with a negative weight. Finally, the events from the background region are used as background.

Figure 3.18 shows the invariant mass distribution of the benchmark decay, after a cut on a BDT trained on *detector data* using Sideband-Subtraction. The exact choice of the regions, as long as they fulfill the assumptions and contain enough statistics, is not important. The different regions were chosen as follows: the signal-enriched region $1.82\,\text{GeV} < M < 1.9\,\text{GeV}$, the background region $1.77\,\text{GeV} < M < 1.78\,\text{GeV}$ and $1.94\,\text{GeV} < M < 1.95\,\text{GeV}$, and the negative signal-region $M < 1.8\,\text{GeV}$ and $1.92\,\text{GeV} < M$. The achieved signal-to-noise ratio is worse than the baseline (see Fig. 3.16).

The method can be used to increase the signal-to-noise ratio in situations were *Monte Carlo events* cannot be used. The expected number of signal events in the signal-enriched region has to be known. If the classifier is able to distinguish background events in the signal and negative signal regions using its features, the classification quality deteriorates, because the negative signal events are partly rejected and cannot subtract all background events in the signal-enriched region. In consequence, only a subset of features can be used during the Sideband-Subtraction training.

3.3.3.4 sPlot

sPlot is a statistical technique, to reconstruct the distribution of different *sources* in so-called *control variables*, using the known distributions of these sources in a *discriminating variable* [26]. In the context of machine learning it can be used to perform a multivariate training on data.

The *sources* are chosen to be signal and background decays, the *control variables* are the features we want to use in the training of the multivariate classifier, and the *discriminating variable* is a variable for which the signal and background distribution is known e.g. the invariant mass of a particle.

[10] Meaning that the events cannot be distinguished by the features used in the training of the classifier.

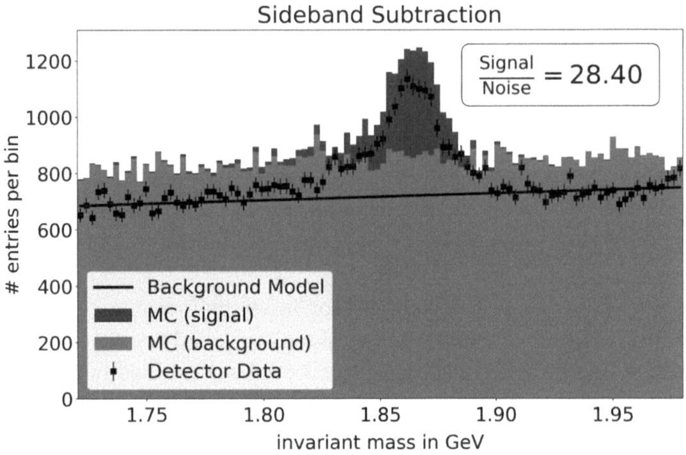

Fig. 3.18 The invariant mass distribution of the benchmark decay $D^0 \to K^-\pi^+\pi^0$, after a cut on the response of a boosted decision tree was applied. The BDT was trained using sideband subtraction. In particular, no *MC events* were used during this training. The filled histograms show the expected distribution from *Monte Carlo* simulation for signal (dark gray) and background (light gray). The black points show the distribution obtained from *detector data*. The black line shows the background model fitted to the *detector data*

The underlying idea is similar to Sideband-Subtraction. The known distributions in the *discriminating variable* define implicitly signal-enriched and background-enriched regions. In contrast to Sideband-Subtraction, sPlot does not require a signal-free region.

In the sPlot based multivariate training every event is used twice, once as signal and once as background, but with different weights. The weights for signal w_s and background w_b are derived from the probability density distributions of the *discriminating variable* for signal f_s and background f_b. These are fitted to the *detector data* in order to extract the number of signal N_s and background events N_b:

$$\begin{pmatrix} w_s \\ w_b \end{pmatrix} = \frac{V}{N_s \cdot f_s + N_b \cdot f_b} \cdot \begin{pmatrix} f_s \\ f_b \end{pmatrix},$$

where V is the covariance matrix of the extracted yields. The approach can be generalized to more than one background source.

sPlot requires the probability density distribution of the *discriminating variable* for signal and all background sources. In addition, the events of each source must be conditional independent of the *discriminating variable*, meaning the classifier should not be able to estimate the value of the *discriminating variable* using its features. In consequence, only a subset of features can be used during the sPlot training. A violation of the above assumptions leads to a deterioration of the achieved classification quality. This is the same effect already encountered in Sect. 3.3.3.3.

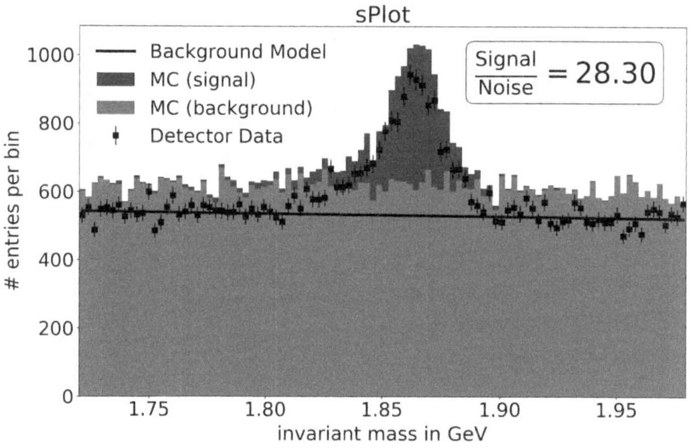

Fig. 3.19 The invariant mass distribution of the benchmark decay $D^0 \rightarrow K^-\pi^+\pi^0$, after a cut on the response of a boosted decision tree was applied. The BDT was trained using sPlot. In particular, no *MC events* were used during this training. The filled histograms show the expected distribution from *Monte Carlo* simulation for signal (dark gray) and background (light gray). The black points show the distribution obtained from *detector data*. The black line shows the background model fitted to the *detector data*

Figure 3.19 shows the invariant mass distribution of the benchmark decay, after a cut on a BDT trained on *detector data* using sPlot. The probability density functions for the invariant mass for signal and background is extracted from Monte Carlo simulation. The achieved signal-to-noise ratio is worse than the baseline (see Fig. 3.16). The more detailed description of the method can be found in [27, 51].

3.3.3.5 Decorrelated sPlot

As stated in the last section, the classification quality of an sPlot training deteriorates if the classifier can predict the *discriminating variable* for background-events using its features. Using the uniformity-constraint is not straight-forward in this case (see Sect. 3.3.3.6).

During this thesis an extension of sPlot was developed to mitigate this problem, by automatically reducing the conditional dependency between the features and the *discriminating variable* for background-events.

The basic idea is to perform an event re-weighting (similar to Sect. 3.3.3.2) to eliminate the undesired conditional dependency. Therefor, a classifier is trained to distinguish events with a low L and high H value of the *discriminating variable*. The main challenge is to leave the dependency between the features and the target (signal or background) untouched.

One possible approach would be to assign all events left of the signal peak to L and all events right of the signal peak to H. In case of a symmetrical signal peak distribution in the *discriminating variable*, there would be the same amount of signal in L and H, in consequence the classifier would learn to distinguish L and H but not signal and background, hence leaving the dependency between the features and the target untouched.

A more advanced approach uses all events twice (similar to sPlot), once as L and once as H, weighted using the cumulative distribution function for signal and normalized to the probability distribution function of background:

$$w_L = \frac{\int_{-\infty}^{x} f_s(x')\mathrm{d}x'}{f_b(x)}$$
$$w_H = \frac{\int_{x}^{\infty} f_s(x')\mathrm{d}x'}{f_b(x)},$$

where x is the *discriminating variable* and f_s (f_b) is the probability density function for signal (background). This approach is independent of assumptions on the signal distribution shape, but still distributes the amount of signal per background evenly between L and H.

In the next step the output p of the previously trained L versus H classifier, is used to assign a decorrelation weight to the event

$$w_D = \frac{1}{2}\left(\frac{\int_{-\infty}^{x} f_s(x')\mathrm{d}x'}{p} + \frac{\int_{x}^{\infty} f_s(x')\mathrm{d}x'}{1-p}\right),$$

in order to remove the conditional dependency between features and the *discriminating variable* for background-events. The cumulative distribution function for signal is again used to leave the signal-events on average untouched. If the classifier could not decide between L and H, that is the probability is close to $p = 0.5$, the final weight is close to $w_D = 1$. In contrast, a stark prediction like a value close to $p = 1$, will lead to a small (large) weight if the prediction is correct (incorrect).

Finally, an sPlot training as described in Sect. 3.3.3.4 is performed using the decorrelation weights w_D multiplied with the sPlot weights.

Figure 3.20 shows the invariant mass distribution of the benchmark decay, after a cut on a BDT trained on *detector data* using decorrelated sPlot. As expected the obtained signal-to-noise ration is better than the one obtained by pure sPlot. Furthermore, the achieved signal-to-noise ratio is not significantly different from the baseline (see Fig. 3.16).

3.3.3.6 Conclusion

All presented methods yield reasonable results on *detector data*. They differ in the amount of information they require from *Monte Carlo* simulation: event re-weighting requires simulated *MC events* to which corrections can be applied; Sideband-Subtraction requires an estimate of the number of signal events in the signal-enriched

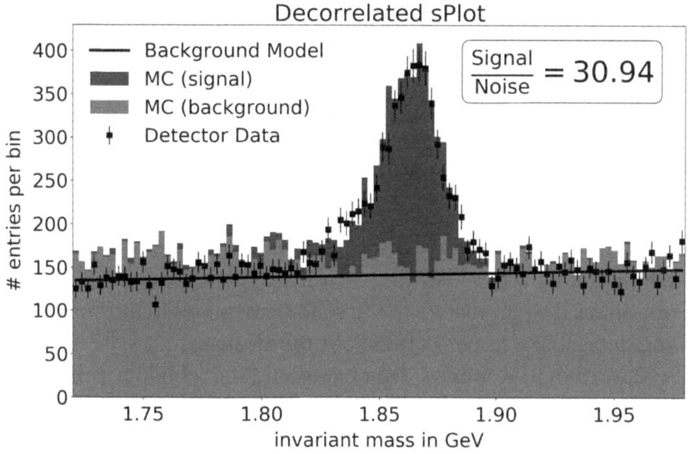

Fig. 3.20 The invariant mass distribution of the benchmark decay $D^0 \rightarrow K^-\pi^+\pi^0$, after a cut on the response of a boosted decision tree was applied. The BDT was trained using decorrelated sPlot. In particular, no *MC events* were used during this training. The filled histograms show the expected distribution from *Monte Carlo* simulation for signal (dark gray) and background (light gray). The black points show the distribution obtained from *detector data*. The black line shows the background model fitted to the *detector data*

region; (decorrelated) sPlot requires the distribution of the signal and background component. In addition, Sideband-Subtraction and sPlot require that the background composition does not change in the considered phase-space region, and that the features used during the training are independent of the *discriminating variable*. The last requirement can be mitigated by the decorrelated sPlot algorithm developed during this thesis.

In consequence, the preferred method depends on the quality of the *Monte Carlo* simulation and the correctness of the necessary assumptions for the investigated tasks.

In the benchmark example presented in this section, the classifier based on pure *Monte Carlo*, still performs slightly better or equally well compared to the data-driven methods. This is expected since the benchmark decay is a well-known and well-simulated decay-channel, except for the continuum component, where the overall normalization is off. The event re-weighting method is particularly useful, because it yields the same results as the pure *Monte Carlo* training, if *MC events* and *detector data* are indistinguishable. It is key to a successful employment of deep learning methods, where low-level features are used, which are usually less well simulated compared to high-level features.

It is not clear yet, if the data-driven techniques presented in this section can be combined with the uniformity-constraint introduced in the last section. While most data-driven technique require a *discriminating-variable* to define signal and background events, the uniformity-constraint is enforced using this definition on the *fit-variable*, which could be the same as the *discriminating-variable* (like it is the case for sPlot). This was not further investigated during this thesis.

3.3.4 Hyper-parameter Optimization

As described in Sect. 3.1.3, the complexity of the statistical model generated by an MVA algorithm can be controlled by so-called hyper-parameters. While the MVA algorithm optimizes the parameters of the statistical model to minimize the error on a given training dataset, the hyper-parameter optimization tunes the hyper-parameters of the MVA algorithm to minimize the error on an independent validation dataset.

In total, three independent datasets are required in this situation: the training dataset used by the MVA algorithm; the validation dataset used by the hyper-parameter optimization; and the test dataset used to evaluate the performance of the selected model. Techniques like cross-validation [1, Sect. 1.3] and Bayesian model comparison [1, Sect. 3.4] can be used to avoid introducing a dedicated validation dataset if the amount of available labeled data is limited.

Optimizing hyper-parameters by hand is cumbersome, error-prone and not suitable for automation. During this thesis different algorithms, used to automatize the search for the optimal hyper-parameter values, were investigated. The `mva` package includes examples for all of them. Figure 3.21 shows the application of the hyper-parameter optimization algorithms, described in the following, to the hyper-parameters of `FastBDT` on the benchmark.

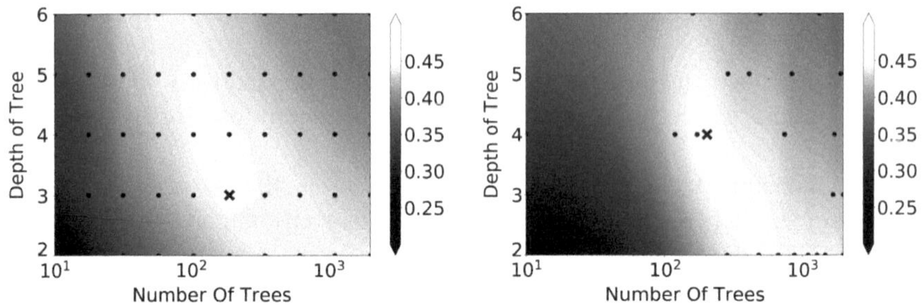

(a) Grid-search evaluated on 50 grid points. (b) Bayesian optimization after 26 evaluations.

Fig. 3.21 Hyper-parameter optimization of `FastBDT` on the benchmark (see Sect. 3.2.2). Two hyper-parameters were optimized: the `number of trees` and the `depth of the trees`. The contour map indicates the ROC AUC score predicted by the measured hyper-parameter settings. The cross indicates the best hyper-parameter configuration with a ROC AUC score of 0.45 (both algorithms found distinct but equivalent maximums)

3.3.4.1 Grid and Random Search

The most common approach is a grid search. The loss on the validation dataset is evaluated on an n-dimensional grid in the hyper-parameter space. The computational effort of this approach scales exponentially in the number of hyper-parameters. The accuracy of the method depends on the density of the grid. Quantitative[11] and qualitative[12] hyper-parameters can be optimized by this approach.

Figure 3.21a shows an application of the grid search algorithm on the benchmark. In total, 50 distinct configurations of `FastBDT` were tested, the best achieved ROC AUC score was 0.450, which is significantly better than the default configuration achieving 0.435 (see Table 3.1).

Recent studies suggest that random-search is superior to grid-search at least in high-dimensional scenarios [12]. This is due to the fact that often only a few out of many hyper-parameters really matter. In such a scenario, the computational effort of random-search is independent of the number of nuisance[13] hyper-parameters.

3.3.4.2 Bayesian Optimization

A probabilistic $f(\vec{h})$ model of the error on the validation dataset given the hyper-parameters $\vec{h}$ is constructed. Using the Bayesian Optimization framework [14] this model can be used to determine the next hyper-parameter space-point to evaluate using all available information from previous evaluations. This approach can take into account prior knowledge about the hyper-parameters and the cost (runtime) of the evaluation for different hyper-parameters. Quantitative and qualitative hyper-parameters can be optimized by this approach.

Figure 3.21 shows an application of the Bayesian optimization algorithm on the benchmark. After 26 evaluations, the model found a configuration with an ROC AUC score of 0.450, which is significantly better than the default configuration achieving 0.435 (see Table 3.1). In comparison to grid-search, Bayesian optimization found an equivalent best hyper-parameter configuration in fewer evaluations.

3.4 Conclusion

In the early years of HEP cut-based analysis (see Sect. 3.1.4.1) dominated the field, in the final years of the last millennium HEP started to adopt multivariate analysis

[11] Quantitative parameters have an intrinsic ordering, examples are the `number of trees` in an BDT, or the `weight-decay` constant in the loss function of a NN.

[12] Qualitative parameters do not have an intrinsic ordering, examples are the separation gain measure in a BDT and the optimization algorithm of a NN.

[13] A hyper-parameter with minor or no influence on the score.

methods and machine learning (see Sect. 3.1.4.2), today the field moves on to deep learning (see Sect. 3.1.4.3).

The `mva` package enables Belle II physicists to keep up with the rapid developments in the field and to easily employ modern machine learning algorithms in their work. Most of the multivariate methods used in the reconstruction and analysis algorithms in `BASF2` are built on the `mva` package and use the default classification method `FastBDT`, both developed during this thesis.

In particular, tagging algorithms, which exploit the unique environment provided by B factories, such as the `Flavour Tagger` and `Full Event Interpretation`, rely heavily on the `mva` package.

The provided algorithms were tested, where appropriate, on data recorded by the Belle experiment by taking advantage of the `b2bii` package.

References

1. C.M. Bishop, *Pattern Recognition and Machine Learning (Information Science and Statistics)* (Springer-Verlag, New York, Inc., 2006). ISBN: 0387310738
2. T. Hastie, R. Tibshirani, J. Friedman, *The Elements of Statistical Learning* (Springer-Verlag, New York, Inc., 2001). ISBN: 978-0-387-84858-7
3. V. Vapnik, Principles of risk minimization for learning theory, in *NIPS* (1991)
4. L. Rosasco et al., Are loss functions all the same? Neural Comput. **16**(5), 1063–1076 (2004). https://doi.org/10.1162/089976604773135104
5. P. McCullagh, J.A. Nelder, *Generalized Linear Models*, 2nd edn. Chapman & Hall (1989). ISBN: 9780412317606
6. E. Parzen, On estimation of a probability density function and mode. Ann. Math. Stat. **33**(3), 1065–1076 (1962). https://doi.org/10.1214/aoms/1177704472
7. R.A. Rigby, D.M. Stasinopoulos, Generalized additive models for location, scale and shape. J. R. Stat. Soc. Ser. C (Appl. Stat.) **54**(3), 507–554 (2005). https://doi.org/10.1111/j.1467-9876.2005.00510.x
8. R.W. Koenker, G. Bassett, Regression quantiles. Econometrica **46**(1), 33–50 (1978)
9. J. Neyman, E.S. Pearson, On the problem of the most efficient tests of statistical hypotheses. Philos. Trans. R. Soc. Lond. Ser. A Contain. Pap. Math. Phys. Character **231**, 289–337 (1933). https://doi.org/10.1098/rsta.1933.0009
10. R.A. Fisher, The use of multiple measurements in taxonomic problems. Ann. Eugen. **7**(7), 179–188 (1936). https://doi.org/10.1111/j.1469-1809.1936.tb02137.x
11. J.H. Friedman, Stochastic gradient boosting. Comput. Stat. Data Anal. **38**(4), 367–378 (2002). https://doi.org/10.1016/S0167-9473(01)00065-2
12. J. Bergstra, Y. Bengio, Random search for hyper-parameter optimization. J. Mach. Learn. Res. **13**, 281–305 (2012)
13. D. Maclaurin, D. Duvenaud, R. Adams, Gradient-based hyperparameter optimization through reversible learning, in *Proceedings of the 32nd International Conference on Machine Learning (ICML-15)*, pp. 2113–2122 (2015), http://jmlr.org/proceedings/papers/v37/maclaurin15.pdf
14. J. Snoek, H. Larochelle, R.P. Adams, Practical Bayesian optimization of machine learning algorithms. Adv. Neural Inf. Process. Syst. **25**, 2951–2959 (2012), arXiv: 1206.2944 [stat.ML]
15. O. Behnke, K. Kroeninger, T. Schoerner-Sadenius, G. Schott, *Data Analysis in High Energy Physics*. Wiley-VCH (2013). ISBN: 9783527410583
16. K. Cranmer, I. Yavin, RECAST: extending the impact of existing analyses. JHEP **04**, 038 (2011). https://doi.org/10.1007/JHEP04(2011)038

17. M. Feindt et al., A hierarchical NeuroBayes-based algorithm for full reconstruction of B mesons at B factories. Nucl. Instrum. Methods **A654**, 432–440 (2011). https://doi.org/10.1016/j.nima.2011.06.008
18. K. Hornik, Approximation capabilities of multilayer feedforward networks. Neural Netw. **4**(2), 251–257 (1991). https://doi.org/10.1016/0893-6080(91)90009-T
19. H.W. Lin, M. Tegmark, D. Rolnick, Why does deep and cheap learning work so well? J. Stat. Phys. (2017). https://doi.org/10.1007/s10955-017-1836-5
20. P. Baldi, P. Sadowski, D. Whiteson, Searching for exotic particles in high-energy physics with deep learning. Nat. Commun. **5**, 4308 (2014). https://doi.org/10.1038/ncomms5308
21. Y. Lecun, Y. Bengio, G. Hinton, Deep learning. Nature **2014**, 436–444 (2015). https://doi.org/10.1038/nature14539
22. I. Goodfellow et al., Generative adversarial nets, in *Advances in Neural Information Processing Systems 27*, pp. 2672–2680. Curran Associates Inc. (2014), http://papers.nips.cc/paper/5423-generative-adversarial-nets.pdf
23. G. Louppe, M. Kagan, K. Cranmer, Learning to pivot with adversarial networks, in *NIPS* (2016), arXiv: 1611.01046 [stat.ME]
24. Y. Bengio, A. Courville, P. Vincent, Representation learning: a review and new perspectives. IEEE Trans. Pattern Anal. Mach. Intell. **35**(8), 1798–1828 (2013). https://doi.org/10.1109/TPAMI.2013.50
25. O. Vinyals, A. Toshev, S. Bengio, D. Erhan, Show and tell: lessons learned from the 2015 MSCOCO image captioning challenge. IEEE Trans. Pattern Anal. Mach. Intell. **39**(4), 652–663 (2017). https://doi.org/10.1109/TPAMI.2016.2587640
26. M. Pivk, F.R. Le Diberder, SPlot: a statistical tool to unfold data distributions. Nucl. Instrum. Methods **A555**, 356–369 (2005). https://doi.org/10.1016/j.nima.2005.08.106
27. D. Martschei, M. Feindt, S. Honc, J. Wagner-Kuhr, Advanced event reweighting using multivariate analysis, in *Proceedings, 14th International Workshop on Advanced Computing and Analysis Techniques in Physics Research (ACAT 2011)*, vol. 368, p. 012028 (2012). https://doi.org/10.1088/1742-6596/368/1/012028
28. T. Keck, FastBDT: a speed-optimized multivariate classification algorithm for the Belle II experiment. Comput. Softw. Big Sci. **1**(1) (2017). https://doi.org/10.1007/s41781-017-0002-8
29. 02 October 2017, https://github.com/thomaskeck/FastBDT
30. J. Therhaag et al., TMVA–Toolkit for multivariate data analysis. AIP Conf. Proc. **1504**(1), 1013–1016 (2012). https://doi.org/10.1063/1.4771869
31. S. Nissen, Implementation of a fast artificial neural network library (FANN). Technical report, Department of Computer Science University of Copenhagen (DIKU) (2003), http://fann.sf.net
32. M. Feindt, U. Kerzel, The NeuroBayes neural network package. Nucl. Instrum. Methods **A559**, 190–194 (2006). https://doi.org/10.1016/j.nima.2005.11.166
33. F. Pedregosa et al., Scikit-learn: machine learning in Python. J. Mach. Learn. Res. **12**, 2825–2830 (2011)
34. T. Chen, C. Guestrin, XGBoost: a scalable tree boosting system, in *Proceedings of the 22nd ACM SIGKDD International Conference on Knowledge Discovery and Data Mining*, pp. 785–794 (2016). https://doi.org/10.1145/2939672.2939785
35. M. Abadi et al., TensorFlow: a system for large-scale machine learning, in *12th USENIX Symposium on Operating Systems Design and Implementation (OSDI 16)*, pp. 265–283 (2016), https://www.usenix.org/system/files/conference/osdi16/osdi16-abadi.pdf
36. F. Chollet et al., Keras (2015), https://github.com/fchollet/keras
37. I.J. Goodfellow et al., Pylearn2: a machine learning research library (2013), arXiv: 1308.4214 [stat.ML]
38. R. Al-Rfou et al., Theano: a Python framework for fast computation of mathematical expressions, arXiv: 1605.02688 [cs.SC]
39. C. Patrignani et al., Review of particle physics. Chin. Phys. **C40**(10), 100001 (2016). https://doi.org/10.1088/1674-1137/40/10/100001

40. J.A. Hanley, B.J. McNeil, The meaning and use of the area under a receiver operating characteristic (ROC) curve. Radiology **143**(1), 29–36 (1982). https://doi.org/10.1148/radiology.143.1.7063747
41. M. Gelb, Neutral B Meson flavor tagging for Belle II. MA thesis, KIT (2015), https://ekp-invenio.physik.uni-karlsruhe.de/record/48719
42. J. Gemmler, Study of B Meson flavor tagging with deep neural networks at Belle and Belle II. MA thesis, KIT (2016), https://ekp-invenio.physik.uni-karlsruhe.de/record/48849
43. D.M. Asner, M. Athanas, D.W. Bliss et al., Search for exclusive charmless hadronic B decays. Phys. Rev. D **53**, 1039–1050 (1996). https://doi.org/10.1103/PhysRevD.53.1039
44. G.C. Fox, S. Wolfram, Observables for the analysis of event shapes in $e^{+}e_{-}$ annihilation and other processes. Phys. Rev. Lett. **41**, 1581–1585 (1978). https://doi.org/10.1103/PhysRevLett.41.1581
45. A.J. Bevan et al., The physics of the B factories. Eur. Phys. J. C **74**, 3026 (2014). https://doi.org/10.1140/epjc/s10052-014-3026-9
46. D. Weyland, Continuum suppression with deep learning techniques for the Belle II experiment. MA thesis, KIT (2017), https://ekp-invenio.physik.uni-karlsruhe.de/record/48934
47. A. Rogozhnikov et al., New approaches for boosting to uniformity. JINST **10**(03), T03002 (2015). https://doi.org/10.1088/1748-0221/10/03/T03002
48. M. Feindt, M. Prim, An algorithm for quantifying dependence in multivariate data sets. Nucl. Instrum. Methods **A698**, 84–89 (2013). https://doi.org/10.1016/j.nima.2012.09.043
49. J. Dolen et al., Thinking outside the ROCs: designing decorrelated taggers (DDT) for jet substructure. JHEP **05**, 156 (2016). https://doi.org/10.1007/JHEP05(2016)156
50. J. Stevens, M. Williams, uBoost: a boosting method for producing uniform selection efficiencies from multivariate classifiers. JINST **8**, P12013 (2013). https://doi.org/10.1088/1748-0221/8/12/P12013
51. B. Lipp, sPlot-based training of multivariate classifiers in the Belle II analysis software framework. BA thesis, KIT (2015), https://ekp-invenio.physik.uni-karlsruhe.de/record/48717

Chapter 4
Full Event Interpretation

The `Full Event Interpretation (FEI)` is a tagging algorithm based on machine learning. It exploits the unique experimental setup of B factory experiments such as the Belle and Belle II experiment. Both experiments operate on the $\Upsilon(4S)$ resonance, which decays at least 96% of the time into exactly two B mesons. Conceptually, the event is divided into two sides: The signal-side containing the tracks and clusters compatible with the assumed signal B_{sig} decay the physicist is interested in, e.g., $B^+ \rightarrow \tau^+\nu_\tau$; and the tag-side containing the remaining tracks and clusters compatible with an arbitrary B_{tag} meson decay. Figure 4.1 depicts this situation. Ideally, a **full** reconstruction of the entire **event** has to take all detected tracks and clusters into account to attain a correct **interpretation** of the measured data.

The `FEI` automatically reconstructs B_{tag} candidates[1] and calculates a signal probability to separate correctly reconstructed B_{tag} candidates from background. Using constraints derived from the unique experimental setup, the reconstructed B_{tag} meson can be used to recover information about the remaining B_{sig} meson in the event.

This enables the measurement of a wide range of decays with a minimum amount of detectable information, like $B^+ \rightarrow \tau^+\nu_\tau$, $B^+ \rightarrow \ell^+\nu\gamma$ and $B \rightarrow K\nu\bar{\nu}$, or no detectable information at all, like in the case of $B^0 \rightarrow \nu\bar{\nu}$, in the final state.

In the following I describe tagging at B factories in general (Sect. 4.1); the `FEI` algorithm itself (Sect. 4.2); the validation of the performance of the `FEI` (Sect. 4.3) using converted Belle detector-data and simulated Belle II Monte Carlo; and finally possible future extensions and use-cases (Sect. 4.4).

[1] A candidate consists of an assumed decay-chain, which happened in the detector and was detected by a fixed set of tracks and clusters.

T. Keck, *Machine Learning at the Belle II Experiment*, Springer Theses,
https://doi.org/10.1007/978-3-319-98249-6_4

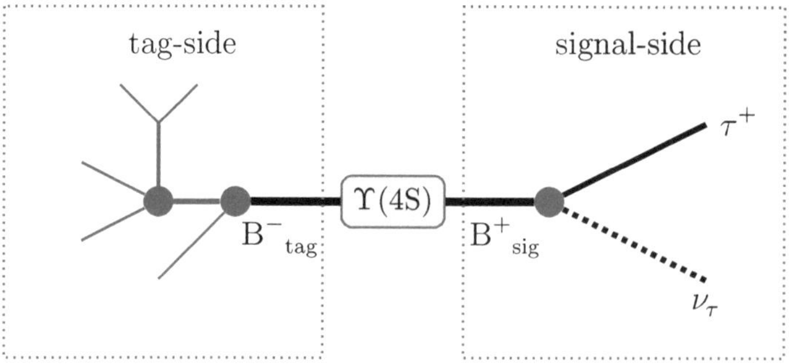

Fig. 4.1 (Left) a common tag-side decay $B^- \to D^0[\to K^0_S[\to \pi^-\pi^+]\pi^-\pi^+]\pi^-$ and (right) the signal-side decay $B^+ \to \tau^+\nu_\tau$ investigated in Chap. 5

4.1 Tag-Side Reconstruction

The initial four-momentum of the produced $\Upsilon(4S)$ resonance is precisely known. Therefore, the reconstruction of the tag-side B meson allows to recover information about the signal-side, which would be otherwise inaccessible. The recovered information includes:

the consistency of quantities like the charge or the flavour;
the four-momentum of the B_{tag} and consequently the B_{sig} meson;
the decay-vertex and the decay time difference Δt between both mesons;
the event-type like $\Upsilon(4S) \to B^0\bar{B}^0$ or $\Upsilon(4S) \to B^+B^-$;
and the assignment of tracks and clusters to either the B_{tag} or the B_{sig} meson.

Historically, there were two distinct approaches to tagging at B factories: inclusive and exclusive.

They differ in their **tagging efficiency** (that is the fraction of $\Upsilon(4S)$ events which can be tagged), their **tag-side efficiency** (that is the fraction of $\Upsilon(4S)$ events with a correct tag) and in the quality of the recovered information, which determines the **purity** (that is the fraction of the tagged $\Upsilon(4S)$ events with a correct tag-side) of the tagged events. These three properties are the key performance indicators used in this thesis. They are closely related to important properties of a specific analysis: The tagging efficiency is important to judge the disk-space required for skimming, that is the number of events which have to be considered for the analysis; the tag-side efficiency influences the effective statistics of the analysis, and the purity is related to the signal-to-noise ratio of the analysis.

Exclusive tagging provides the assignment of tracks and clusters to either the tag-side and or the signal-side, whereas the inclusive tag requires this assignment as input. The advantages and disadvantages of the different approaches are visualized in Fig. 4.2.

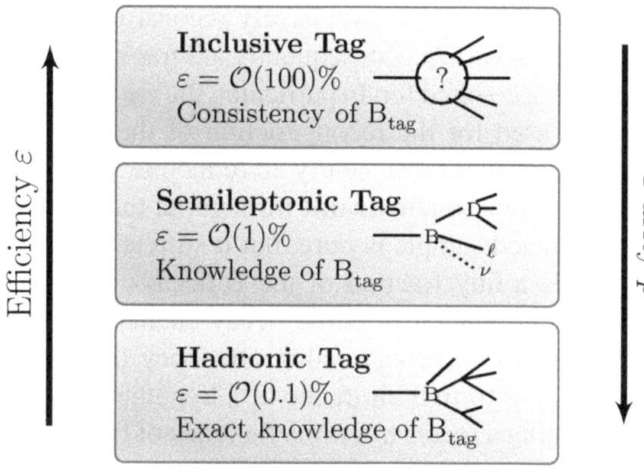

Fig. 4.2 Overview of the three tagging algorithms, which were used in the past to infer information about the B_{sig} meson using the other B_{tag} meson in the event

4.1.1 Inclusive Tagging

Inclusive refers to the reconstruction of a particle (here the B_{tag}) without assuming an explicit decay-channel. Consequently, inclusive tagging combines the four-momenta of all tracks and clusters, which were not used during the reconstruction of the B_{sig}—the so-called rest of event. The decay-chain of the B_{tag} is not explicitly reconstructed, hence the assignment is not provided by the algorithm and cannot be used to discard wrong candidates. Only the overall consistency between the B_{sig} and B_{tag} meson can be checked. This approach has a high tagging efficiency of $\mathcal{O}(100\%)$, since it can always provide a valid B_{tag}, but suffers from a high background, and consequently the tagged sample is very impure.

Inclusive tagging is used in time-dependent CP violation analyses, to determine the decay vertex of the B_{tag} meson. It can also be used to improve the momentum resolution on the signal-side in analyses like $B^+ \to \mu^+ \nu$, where the signal-side alone is suffice to provide a pure sample. Flavour tagging (see Sect. 3.3.1.2) and continuum suppression (see Sect. 3.3.1.3) can be seen as a special form of inclusive tagging. On the other hand, inclusive tagging cannot be used in conjunction with an inclusive signal-side.

The `FEI` is not an inclusive tagging algorithm. However, it does try to always provide a valid B_{tag} like it is accomplished by the inclusive tagging. Therefore, it could be used instead of inclusive tagging as described in Sect. 4.4.1.

4.1.2 Exclusive Tagging

Exclusive refers to the reconstruction of a particle (here the B_{tag}) assuming an explicit decay-channel. Consequently, exclusive tagging reconstructs the B_{tag} independently of the B_{sig} using either hadronic or semileptonic B meson decay-channels. The decay-chain of the B_{tag} is explicitly reconstructed and therefore the assignment of tracks and clusters to the tag-side and signal-side is known.

If the signal-side is exclusively reconstructed as well, the entire decay-chain of the $\Upsilon(4S)$ is known. Consequently, all tracks and clusters measured by the detector should be accounted for. In particular, the requirement of no additional tracks, besides the ones used for the reconstruction of the $\Upsilon(4S)$, is an extremely powerful and efficient way to remove nearly all reducible[2] background. This requirement is called the **completeness-constraint** throughout the text.

The tagged sample is pure, but it suffers from a low tag-side efficiency $\mathcal{O}(1)\%$, since only a tiny fraction of the B decays can be explicitly reconstructed, due to the large amount of possible decay-channels and their high multiplicity, as well as the imperfect reconstruction efficiency of tracks and clusters. For instance, the probability p to find all tracks of a hypothetical decay chain with a multiplicity of 10, assuming a reconstruction efficiency of 0.95 per track and a geometrical detector acceptance of 91.1% is only $p \approx 24\%$.

The quality of the recovered information and systematic uncertainties depend on the decay-channel of the B_{tag}, therefore one distinguishes further between hadronic and semileptonic exclusive tagging.

Additionally, exclusive tagging can also be used to perform an inclusive reconstruction of the signal-side, without assuming an exclusive signal decay-channel.

The `FEI` is an exclusive tagging algorithm, which supports hadronic and semileptonic tagging (Fig. 4.3).

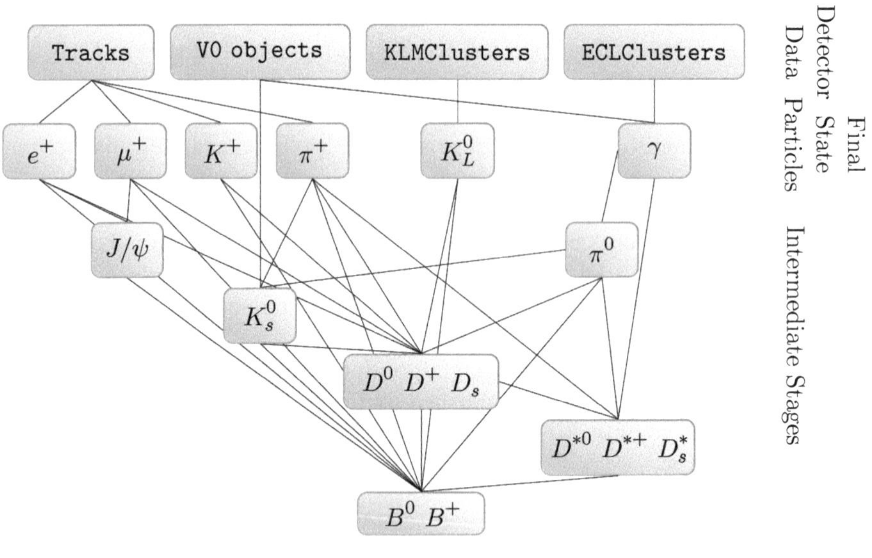

Fig. 4.3 Schematic overview of the `FEI`. The B mesons are reconstructed hierarchically; using the detector output, final state particle candidates are formed, and combined to intermediate particles until the final B candidates are formed. The probability of each candidate to be correct is estimated by an MVA algorithm. The probabilities are fed into the subsequent MVA methods

[2]Reducible background has distinct final state products from the signal.

4.1.2.1 Hadronic Tagging

Hadronic tagging solely uses hadronic B decay-channels for the reconstruction. Hence, the four-momentum of the reconstructed B_{tag} is well-known and the tagged sample is very pure. However, hadronic tagging suffers from a low tag-side efficiency. It is only possible for a tiny fraction of the recorded events, because the branching fraction of explicit fully hadronic decay-chains is very small. A typical hadronic B decay has a branching fraction of $\mathcal{O}(10^{-3})$.

Figures 4.5 and 4.10 show the beam-constrained mass distribution of hadronically reconstructed B_{tag} mesons for Belle and Belle II, respectively. The correctly reconstructed mesons peak at a beam-constrained mass of $M_{bc} = 5.279\,\text{GeV}$. The beam-constrained mass can be calculated from the initial beam-conditions [1, Chap. 7.1]:

$$M_{bc} = \sqrt{E_{beam}^2 - p_B^2}, \tag{4.1}$$

where all quantities are measured in the center-of-mass system and E_{beam} is the beam energy. M_{bc} is per definition independent of the mass hypotheses of the final state particles used during the reconstruction of the B meson candidate.

Another useful quantity is the deviation from the beam-energy [1, Chap. 7.1]

$$\Delta E = E_B - E_{beam}, \tag{4.2}$$

which is sensitive to mis-identification of the final state particles during the reconstruction of the B meson candidate.

The above beam-dependent quantities are less correlated with each other in comparison to the invariant mass M and energy E of the reconstructed B meson candidate.

The hadronic tag of the `FEI` can be accessed by the user using `BASF2` via the provided `B0:generic` and `B+:generic ParticleLists`. The different decay-channels can be distinguished by the associated `decayModeID`.

4.1.2.2 Semileptonic Tagging

Semileptonic tagging uses semileptonic B decay-channels. Due to the clean signature of semileptonic B decays with high branching fraction, this approach usually has a higher tag-side efficiency compared to hadronic tagging. A typical semileptonic B decay has a branching fraction of $\mathcal{O}(10^{-2})$. On the other hand, the semileptonic reconstruction suffers from missing kinematic information due to the neutrino in the final state of the decay. Hence, the sample is not as pure as in the hadronic case.

Figures 4.6 and 4.11 show the angle $\cos\Theta_{BD\ell}$ between the true B meson and the measured $D\ell$ system of semileptonically reconstructed B_{tag} mesons for Belle and Belle II, respectively. The correctly reconstructed mesons are missing exactly one massless neutrino, and therefore peak between -1 and 1. The angle $\cos\Theta_{BD\ell}$ can be calculated from the initial beam-conditions [1, Chap. 7.2]:

$$\cos\Theta_{\mathrm{BD}\ell} = \frac{2E_{\mathrm{beam}}E_{\mathrm{D}\ell} - m_{\mathrm{B}}^2 - m_{\mathrm{D}\ell}^2}{2p_{\mathrm{B}}p_{\mathrm{D}\ell}}, \tag{4.3}$$

where all quantities are measured in the center-of-mass system and E_{beam} is the beam energy, m_{B} is the nominal mass of the B meson and p_{B} is the total momentum of a B meson from a $\Upsilon(4S)$ decay.

The semileptonic tag of the `FEI` can be accessed by the user using `BASF2` via the provided `B0:semileptonic` and `B+:semileptonic` `ParticleLists`. The different decay-channels can be distinguished by the associated `decayModeID`.

4.2 Implementation

A proof of concepts (PoC) of the `Full Event Interpretation` was developed in [2] based on the original `Full Reconstruction (FR)` algorithm used by Belle [3]. During this thesis, the `FEI` was further developed from a PoC to a standard tool used in production. It was significantly extended by: further increasing the tagging and tag-side efficiency by getting more inclusive; reducing the runtime and memory consumption; adapting the algorithm for the usage on converted Belle data using `b2bii`; and migrating the algorithm to the `mva` package.

In the following I describe: the complete `FEI` algorithm in an abstract manner (Sect. 4.2.1), the training of the employed multivariate classifiers (Sect. 4.2.2), and the performance optimizations (Sect. 4.2.4) which are key to a successful employment of the `FEI` in the future.

4.2.1 Algorithm

The FEI follows a hierarchical approach with six stages, visualized in Fig. 4.3. All steps in the algorithm are configurable, therefore: the used decay-channels, employed cuts, and the input features and hyper-parameters of the multivariate classifiers depend on the configuration. A detailed description of the current default configuration of the `FEI` including a list of all decay-channels can be found in Sect. B.1.

4.2.1.1 Combination of Candidates

Charged final state particle candidates are created from `Tracks` assuming different particle hypotheses, whereas neutral final state particle candidates are created from `ECL clusters`, `KLM clusters`, or `V0 objects`. Each candidate can be correct (signal) or wrong (background). For instance, a `Track` used to create a π^+ candidate can originate from a pion traversing the detector (signal), from a kaon traversing the detector (background) or even consists of a random combination of hits from beam-background (also background).

All candidates available at the current stage are combined to intermediate particle candidates in the subsequent stages, until candidates for the desired B mesons are created. Each intermediate particle has multiple possible decay-channels, which can be used to create valid candidates. For instance, a B^- candidate can be reconstructed by combining a D^0 and a π^- candidate, or by combining a D^0, a π^- and a π^0 candidate. The used D^0 candidate could be reconstructed from a K^- and a τ^+, or from a K_S^0 and a π^0.

The `FEI` reconstructs more than $\mathcal{O}(100)$ explicit decay-channels, leading to more than $\mathcal{O}(10000)$ distinct decay-chains. All considered decay-channels are listed in Sect. B.1.

4.2.1.2 Multivariate Classification

The `FEI` employs multivariate classifiers (see Chap. 3) to estimate the probability of each candidate to be correct. Hence, each candidate created by the `FEI` (regardless at which stage) has an associated `SignalProbability` σ, which can be used to discriminate correctly reconstructed candidates from background.

In order to use all available information, a network of multivariate classifiers is built, following the hierarchical structure of the reconstruction. For each final state particle and for each decay-channel of an intermediate particle, a multivariate classifier is trained which estimates the probability that the candidate is correct.

For instance, the classifier built for the decay of $B^- \rightarrow D^0\pi^-$ would use the `SignalProbabilitys` of the used D^0 and π^- candidates, to estimate the `SignalProbability` of the B^- candidate created by combining the aforementioned D^0 and π^- candidates.

The input features of the classifier are among others the kinematics and vertex fitting information of the candidate and its daughters, as well as the `SignalProbability` of the daughters. The full list of input features and the chosen hyper-parameters for the multivariate classifiers can be found in Sect. B.1.

As can be seen in Fig. 4.3 the available information flows from the data provided by the detector, through the intermediate candidates into the final B meson candidates, yielding a single number, which can be used to distinguish correctly reconstructed from incorrectly reconstructed B_{tag} mesons.

4.2.1.3 Combinatorics

It is not possible to consider all possible B meson candidates created by all possible combinations. The amount of possible combinations scales with the factorial in the number of tracks and clusters. This problem is known as **combinatorics** in HEP. Furthermore, it is not worthwhile to consider all possible B meson candidates, because all of them (except for two in the best-case scenario) are wrong.

The `FEI` uses two sets of so-called **cuts**. A cut is a criterion a candidate has to fulfill to be further considered. For instance one could demand that the invariant mass of the B meson candidate is near the nominal mass 5.28 GeV of a B meson particle, or that a μ^+ candidate has a large μ likelihood calculated from the measurements in the PID sub-detectors.

Directly after the creation of the candidate (either from a detector object, or by combining other candidates), but before the application of the multivariate method, the FEI uses loose and fast **pre-cuts** to remove wrongly reconstructed candidates (background), without loosing signal. The main purpose of these cuts is to save computing time and to reduce the memory consumption. These **pre-cuts** are applied separately for each decay-channel.

At first, a very loose fixed cut is applied on a quantity, which is fast to calculate, e.g. the energy for photons, the invariant mass for D mesons, the released energy in the decay for D* mesons, or the beam-constrained mass for hadronic B mesons.

Secondly, the remaining candidates are ranked according to a quantity, which is fast to calculate (usually the same quantity as above is used here). Only the n best-candidates in each decay-channel are further considered, the others are discarded. This best-candidate selection ensures that each decay-channel and each event receives roughly the same amount of computing time.

Next, the computationally expensive parts of the reconstruction are performed on each candidate (see also Sect. 4.2.4): the Monte Carlo matching (in case of MC), the vertex fitting, and the multivariate classification.

After the multivariate classifiers have estimated the `SignalProbability` of each candidate, the candidates of different decay-channels can be compared to one another. Here the FEI uses tighter **post-cuts** to aggressively remove wrongly reconstructed candidates using all available information. The main purpose of these cuts is to restrict the number of candidates per particle to a manageable number.

At first, there is a loose fixed cut on the `SignalProbability`, to remove unreasonable candidates.

Secondly, the remaining candidates are ranked according to their `SignalProbability`. Only the m best-candidates of the particle (that is over all decay-channels) are further considered, the others are discarded. This best-candidate selection ensures that the amount of candidates produced in the next stage is tractable by the computing system (see Sect. 4.2.4.1).

The exact definitions of the **pre-cuts** and **post-cuts** like the used values for n and m can be found in Sect. B.1.

4.2.2 Training

The multivariate classifiers used by the FEI are trained on Monte Carlo (MC) simulated events as explained in Chap. 3. Each multivariate classifier requires at least $\mathcal{O}(10^3)$ signal and background candidates to be successfully trained, that is to prevent under-fitting and over-fitting. The branching fractions of the employed decay-channels for B and D mesons are $\mathcal{O}(10^{-3})$ and the reconstruction efficiency of some B decay-channels is as low as $\mathcal{O}(10^{-2})$. So, it is expected to require $\mathcal{O}(10^8)$ MC events containing B meson decays for a successful training of the FEI. This number was empirically confirmed.

During the training of the FEI, the provided MC is reconstructed following the hierarchical approach. At each stage the necessary training data for all classifiers used by the current stage is written out, and the reconstruction is suspended before the **post-cuts** would have been applied. The classifiers are fitted using the training data. Afterwards, the reconstruction is resumed, the **post-cuts** are now applied using the immediately previously fitted classifiers. The technical aspects of the training are discussed in Sect. 4.2.4.4.

There are three distinct types of MC events, which could be used for the training of the `FEI`: **double-generic events**: $e^+e^- \rightarrow \Upsilon(4S) \rightarrow B\bar{B}$, where both B mesons decay generically[3]; **continuum events**, that is non-resonant interactions $e^+e^- \rightarrow q\bar{q}, \ell\bar{\ell}, \gamma\gamma$; and **signal events**: $e^+e^- \rightarrow \Upsilon(4S) \rightarrow B\bar{B}$, where one B decays generically, and the other decays in an analysis-specific signal-channel like $B^+ \rightarrow \tau^+\nu_\tau$.

Depending on the training procedure and mixture of MC types in the training, the multivariate classifiers of the `FEI` are optimized for different objectives. The main objectives of the `FEI` are: a **high tag-side efficiency on signal events**, that is to output correctly reconstructed B_{tag} mesons for signal events and assign them a high `SignalProbability`; and a **high purity** by rejecting background candidates, or assigning them a low `SignalProbability`. Moreover, if the signal-side can already provide a good B_{sig} candidate, it is also desired to obtain a **high tagging efficiency** by always providing at least one reasonable candidate. This can improve the final signal selection efficiency, because the event is not discarded if for instance mis-identified or missing tracks prevent an entirely correct reconstruction of the tag-side.

Continuum events are not used during the training of the `FEI`. Firstly, continuum events do not contain usable signal candidates. B mesons are not present in continuum events, and other particles like D mesons have different properties (e.g. a higher momentum) than typically found in $\Upsilon(4S)$ events. Secondly, they can be suppressed efficiently by the dedicated `ContinuumSuppression` (Sect. 3.3.1.3) algorithm of `BASF2`.

In the following I describe the main training procedures and discuss their respective advantages and disadvantages.

4.2.2.1 Generic FEI

The generic `FEI` is trained on double-generic MC events. The training is done independently of any specific signal-side and is performed centrally, once per Monte Carlo campaign.

The classifiers are trained to identify correctly reconstructed B_{tag} mesons on double-generic events. The training of the classifiers could be sub-optimal for signal events, due to the different multiplicity distribution of double-generic and signal events. One can mitigate this problem by training the `FEI` on double-generic MC events with a maximum number of $12 + N$ tracks, where 12 is the maximum number of tracks the `FEI` can combine to a **valid**[4] B_{tag} and N is the number of tracks required for the reconstruction of the B_{sig} (see Sect. 4.3.3.1).

After the user performed his analysis-specific signal-side selection and applied the completeness-constraint, most of the wrongly reconstructed B_{tag} candidates can

---

[3] The meson decays into all possible final states with the correct branching fractions.

[4] Empirically, 7 is the maximum number of tracks the `FEI` can combine to a **correct** candidate.

be rejected. The generic FEI cannot take advantage of this fact during the training of the classifiers.

4.2.2.2 Specific FEI

The specific FEI is trained evenly on double-generic and signal MC events. First the analysis-specific signal-side selection is performed, afterwards the FEI is trained on the rest-of-event of the B_{sig} candidates.

The classifiers are specifically trained to identify correctly reconstructed B_{tag} mesons for signal events, which is the main objective of the FEI. Furthermore, B classifiers are only trained on those background candidates from the rest-of-event of double-generic events, which fulfill the completeness-constraint. In consequence, the classifiers can focus on reducing the non-trivial background. Candidates from incorrectly reconstructed B_{sig} from signal events are not used at all, because the candidates produced by these events would be vastly over-represented in the training data, since the signal events are usually very rare on data.

Since only B candidates which fulfill the completeness-constraint are used, the procedure suffers from notoriously low statistics in signal and background. In consequence, large amounts of double-generic and signal MC are required to properly train the FEI using this procedure.

The training depends on the analysis-specific signal-side selection and has to be performed by the user for their analysis.

4.2.2.3 Mono FEI

The mono FEI is trained on signal MC events for the decay $B^0 \to \nu\bar{\nu}$. There is exactly one detectable B in each event. In effect, the FEI is trained on the rest-of-event of an arbitrary correctly reconstructed B_{sig}.

The classifiers are specifically trained to identify the most likely tag-side decay-chain, under the assumption of an correctly reconstructed signal-side. Such a training could be used instead of the traditional inclusive tagging algorithm. It can provide improved tag-side vertexing by exploiting the explicitly known tag-side decay-chain, e.g. using the decay-tree fitter (see [4]). This type of training was not further investigated during this thesis.

The training can be done independently of any signal-side and can be performed centrally, once per Monte Carlo campaign.

4.2.3 Application

The FEI can either be applied to the **entire event**, or only to a subset like the **rest-of-event** of a B_{sig} candidate.

After the application the FEI provides tag-side B candidates for B^+ and B^0 in hadronic and semileptonic decay-channels. Each candidate has an associated SignalProbability that can be used to reject wrongly reconstructed candidates.

An analysis with an inclusive signal-side such as $B^+ \rightarrow X_u \ell^- \nu$ applies the generic FEI to the entire event without using the completeness-constraint. The rest-of-event of the B_{tag} candidate can be used to reconstruct the signal-side inclusively by assigning all remaining tracks and clusters to the signal-side.

An analysis with an exclusive signal-side, e.g. $B^+ \rightarrow \tau^+ \nu_\tau$, can either apply the generic FEI to the entire event or the specific FEI to the rest-of-event of the B_{sig}. In the end, the completeness-constraint is always applied. The choice of the generic or specific FEI depends on the feasibility of a dedicated training of the specific FEI using large amounts of signal and double-generic MC events. Furthermore, the systematic uncertainties are different (see Sect. 4.3.4).

Finally, it is also possible to apply the generic FEI to the rest-of-event, although it was trained on the entire event. This allows a rough estimation of improvements which can be expected of the specific FEI.

4.2.4 Performance

Fitting the FEI on $\mathcal{O}(100)$ million events and applying it to $\mathcal{O}(1)$ billion events, is a CPU-intensive task. An optimized runtime and a small memory-footprint is key for the practicalness in production and saves computing resources i.e. money. As can be seen from Table 4.1 most of the time in the FEI application is spent in: vertex fitting, particle combination and classifier inference. All three tasks have been carefully optimized during this thesis. These optimizations are discussed in the following sections.

Furthermore, the time-consuming training procedure has been implemented using a distributable map-reduce approach (Sect. 4.2.4.4).

All runtime measurements in this section were performed on the KEKCC cluster. The measured values depend on the hardware (e.g., CPU, RAM, and caches), the load of the system (e.g., the amount of rivaling processes and the current IO traffic), the software (the exact BASF2 version) and the data (e.g., the Belle experiment and

Table 4.1 Relative time in percent spent by the FEI applied to $\Upsilon(4S)$ Belle II MC for various tasks. The fitting of the FEI is dominated by reading (writing) the cached DataStore from (to) disk space in between the stages, which is limited by the IO throughput of the system and cannot be optimized in the FEI. The application is still dominated by the vertex fitting, which originally required about 78% of the CPU time

Task	Training	Application
Read/Write DataStore	30	0
Vertex fitting	26	38
Particle combination	19	27
Classifier inference	11	15
Training data & monitoring	65	0
Best candidate selection	3	6
Other	5	14

event type). Therefore, only the relative differences between the stated numbers are meaningful.

4.2.4.1 Combination of Candidates

As explained in Sect. 4.2.1.3, the number of candidates which have to be processed is growing like the factorial of the multiplicity of the channel. In previous approaches the runtime and the maximum memory consumption was dominated by a few high multiplicity events and tight cuts had to be applied to high multiplicity channels.

In contrast, the `FEI` limits the combinatorics problem by performing best-candidate selections during the reconstruction of the decay-chain instead of fixed cuts.

In consequence, each event and each decay-channel is allowed to process the same number of candidates in vertex fitting and classifier inference i.e. consuming similar amounts of CPU time. Moreover, the maximum memory consumption is limited due to the fixed number of best-candidates per event, which is a key requirement by the computing infrastructure.

4.2.4.2 Vertex Fitting

The default vertex fitting implementation of Belle II was ported from the legacy Belle code and is named `KFitter` (Kinematic Fitter). It is based on a Kalman Filter and supports unconstrained, IP profile constrained, and mass-constrained fits. In addition, the `RAVE` (Reconstruction in an Abstract, Versatile Environment) toolkit [5] was integrated in `BASF2`. It has its roots in the CMS (Compact Muon Solenoid) vertex reconstruction software and supports a broad collection of vertex finding and fitting algorithms based on a Kalman Filter approach generalized to include adaptive track assignment and multiple vertices.

On the other hand, the `FEI` uses only a simple unconstrained vertex fit during the reconstruction, and feeds the calculated information into its multivariate classifiers. The user can refit the whole decay-chain of the final B candidates again, including mass and/or ip profile constraints if desired. Therefore, the `FEI` restricts itself to a fast unconstrained vertex fitting. Still, the major part of the CPU time of the `FEI` is spent during vertex fitting.

During this thesis a dedicated fitter (called `FastFit`) based on a Kalman Filter [6] was implemented for the `FEI`, which outperforms the default `KFitter` implementation by one order of magnitude as can be seen in Fig. 4.4. A speed-up of the vertex-fitting part of $s_{\text{FastFit}} \approx 5$ of the `FEI` was measured on generic B^+B^- Belle II events. With `KFitter` the application phase of the `FEI` spent $p \approx 78\%$ of the CPU time during the vertex fitting. According to Amdahl's law [7], a theoretical speed-up of the entire `FEI` of approximately

$$S_{\text{FEI}} = \frac{1}{(1-p) + \frac{p}{s_{\text{FastFit}}}} = 2.66,$$

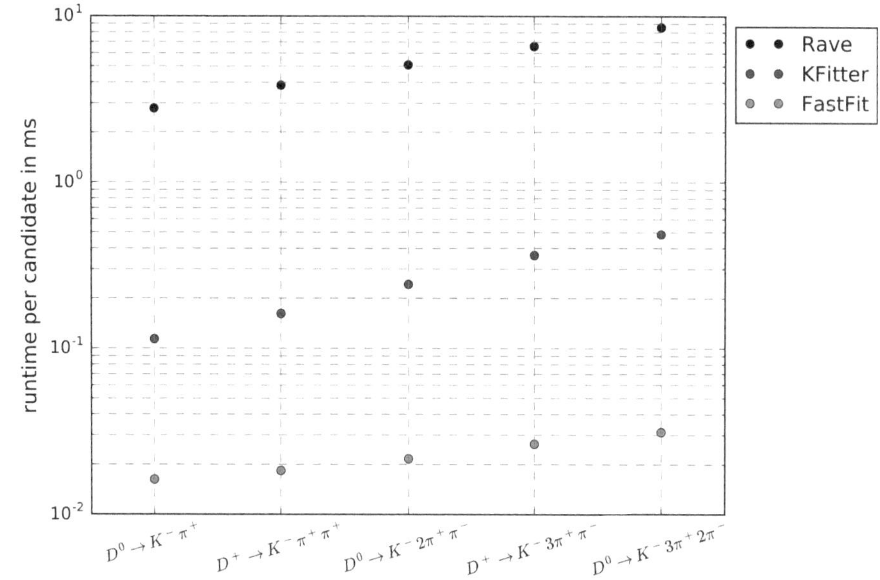

Fig. 4.4 Runtime of an unconstrained vertex fit for different D meson decay-channels in the BASF2 framework. FastFit is one order of magnitude faster than KFitter and two orders of magnitude faster than RAVE

is expected. Empirically, an overall speed-up of the FEI of 2.74 was observed.

The quality (deviation to the true vertex position) of the vertex fit of all three tested implementations is very similar, which is expected since all implementations calculate the same well-defined least-square solution. Deviations occur due to different stopping criterion of the algorithms. Consistently, the overall maximum tag-side efficiency of the FEI does not depend on the employed vertex fitter during fitting or application.

The FastFit code is licensed under GPLv3 and available on github [8].

4.2.4.3 Multivariate Classification

As described in Sect. 3.2.1.1, FastBDT is the default multivariate classification algorithm of BASF2. It was originally designed for the FEI to speed up the fitting and inference-phase. Compared to other popular BDT implementations such as those provided by TMVA, SKLearn and XGBoost it originally gained more than one order of magnitude in execution time, both in fitting and inference. Using FastBDT, most of the time is spent during the extraction of the necessary features used for the inference, therefore no further significant speedups can be achieved by employing a different method.

4.2.4.4 Distributed Training

The training of the FEI is fully automatized and distributed. The distributed training was already introduced in [2]. It follows a map-reduce approach.

The supplied Monte Carlo files are partitioned and processed by independent computing nodes. At first the total amount of supplied Monte Carlo events and their particle content (e.g. the total number of B^+ and B^0 mesons) is calculated. Subsequently, the six stages of the FEI hierarchy are reconstructed.

At each stage the necessary training data for all classifiers used by the current stage is written out. The reconstruction is suspended and the current content of the DataStore of all events on all nodes is cached on disk.

The training data is merged centrally and the associated classifiers are fitted. The generated WeightFiles are uploaded into the local Belle II Conditions Database.

Afterwards, reconstruction is resumed by reading in the cached DataStore files from disk. The classifiers are loaded from the corresponding WeightFiles, which are downloaded from the Belle II Conditions Database.

4.2.4.5 Comparison with Full Reconstruction

The FEI was compared to the hadronic Full Reconstruction (FR) algorithm used by Belle.

Out-of-the-box the FR requires on average 36 ms (39 ms) to reconstruct a simulated $B^0\bar{B}^0$ (B^+B^-) Belle event. The FR was implemented using dedicated C++ code. In contrast, the FEI requires with the identical configuration[5] on average 24 ms (26 ms) to reconstruct the same Belle events using the b2bii interface. The FEI is implemented in Python and only uses general purpose modules implemented in C++.

The default configuration of the FEI requires 98 ms (100 ms) to reconstruct a simulated $B^0\bar{B}^0$ (B^+B^-) Belle event. This time includes the reconstruction of hadronic and semileptonic B_{tag} candidates using many additional channels (see Sect. B.1) and yielding a higher tag-side efficiency.

The average runtime 119 ms (128 ms) to reconstruct a simulated $B^0\bar{B}^0$ (B^+B^-) Belle II event is significantly larger because of increased event size in terms of number of reconstructed tracks and number of reconstructed clusters (Table 4.2)

4.2.5 Automatic Reporting

The FEI includes an automatic reporting system called Full Event Interpretation Report (FEIR). The FEIR contains efficiencies and puri-

[5] Only decay-channels were reconstructed, which were used by the FR as well.

Table 4.2 Application time of the FEI in ms for various types of MC7 on the KEKCC cluster. The uncertainty of the absolute values is about ±3 ms

MC Type	Time per event in ms
$\Upsilon(4S) \to B^+B^-$	128
$\Upsilon(4S) \to B^0\bar{B}^0$	119
$e^+e^- \to c\bar{c}$	76
$e^+e^- \to s\bar{s}$	48
$e^+e^- \to d\bar{d}$	45
$e^+e^- \to u\bar{u}$	43
$e^+e^- \to \tau^-\tau^+$	14
$e^+e^- \to \mu^-\mu^+$	12
$B^+ \to \tau^+\nu_\tau$	43

ties for all particles and decay-channels at different points during the reconstruction. It is generated from so-called monitoring histograms, which can optionally be written out during the training and the application. In addition, reports for each multivariate classifier can be created by the `mva` package (see Sect. 3.2.4).

This built-in monitoring capability upgrades the `FEI` from a black-box to a white-box algorithm, which the user can understand and inspect on all levels.

An excerpt of the `FEIR` can be found in Sect. B.2.

4.3 Validation

The `FEI` was extensively validated on converted Belle MC from the last Belle Monte Carlo campaign, converted Belle data recorded by the Belle detector, and Belle II MC from the 7th Belle II Monte Carlo campaign. Unless stated otherwise all studies were conducted using $\Upsilon(4S)$ and continuum events (including $c\bar{c}$, $s\bar{s}$, $d\bar{d}$, $u\bar{u}$ and $\tau^-\tau^+$) scaled to their expected fractions on recorded data.

The validation includes a detailed discussion on the signal-definition used for the B_{tag} candidates (Sect. 4.3.1); the performance of the `FEI` on converted Belle events (Sect. 4.3.2) and Belle II events (Sect. 4.3.3); the calibration of the `FEI` on converted Belle data using control channels (Sect. 4.3.4); and a comparison to the `Full Reconstruction (FR)` algorithm used by Belle (Sect. 4.2.4.5).

4.3.1 Signal Definitions

This thesis uses the Monte Carlo matching algorithm of `BASF2`. The algorithm searches for the common mother of all tracks and clusters used during the reconstruction. It can distinguish a large number of possible reconstruction errors, which allows for a fine-grained signal-definition:

- particles which were missed during the reconstruction: final state radiation (fsr), photons (excluding fsr), intermediate resonances; neutrinos, K_L^0 or other massive particles;
- charged final state particles, which were misidentified or decayed in flight;
- particles which were wrongly added during the reconstruction.

The **default signal-definition** of Belle II allows for missing final state radiation, missing intermediate resonances and missing neutrinos. If not stated otherwise, this thesis uses this definition, to be comparable to previous exclusive tagging algorithms.

However, since the B mesons reconstructed by the `FEI` are used for exclusive tagging, an **extended signal-definition** can be useful. Here, the mis-identification of a charged final state particle and the addition of beam-induced background particles is allowed as well. The completeness-constraint is uncompromised by this extended signal-definition. Also important quantities like the beam-constrained mass and deviation from the nominal beam-energy are often unchanged.

As can be seen from Figs. 4.5 (Belle) and 4.10 (Belle II), the hadronically tagged extended-signal candidates peak at the same position in the beam-constrained mass as the default-signal candidates. The same holds true in $\cos\Theta_{\mathrm{BD}\ell}$ for the semileptonically tagged B mesons (see Figs. 4.6 (Belle) and 4.11 (Belle II)).

The multivariate classifiers of the `FEI` use the default signal-definition, with an additional constraint for final state particles, which have to come from the primary interaction to be considered signal (for instance electrons from converted gammas are not considered signal in the electron classifier). Therefore the candidates fulfilling only the extended signal-definition are suppressed if a tight cut on the `SignalProbability` is chosen (see Figs. 4.7a and 4.12a).

Even the extended signal-definition does not include all peaking background contributions from mis-reconstructed B mesons, as can be seen from Fig. 4.8, where a Gaussian distribution for the peaking B candidates and an ARGUS function for combinatorial background [1, Chap. 7.1] was fitted to converted Belle data. This additional **peaking background** is caused by nearly correctly reconstructed B mesons. There is no peaking background component visible on recorded off-resonance data, which does not contain B mesons (see Fig. 4.8). Consequently, the peaking background is caused by physics and not by a non-uniform selection efficiency of the classifiers used by the `FEI` (see Sect. 3.3.2). It is debatable if those peaking B candidates can be (for some analyses) considered signal as well.

An analysis, for instance the benchmark analysis $B^+ \to \tau^+\nu_\tau$ (see Chap. 5), usually accepts all events which contain the searched signal-decay as signal. Hence, the signal-definition of the tag-side is not used. Still, many analyses could profit from the relaxed extended signal-definition if the classifiers used by the `FEI` are explicitly trained with this definition.

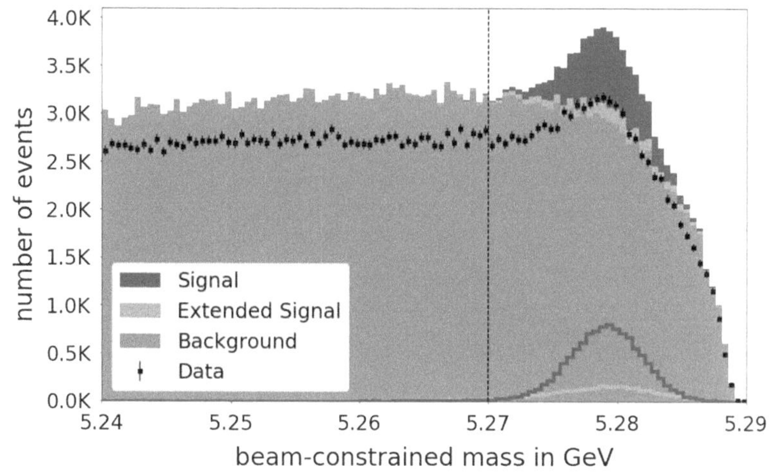

(a) A loose cut on the `SignalProbability` was performed.

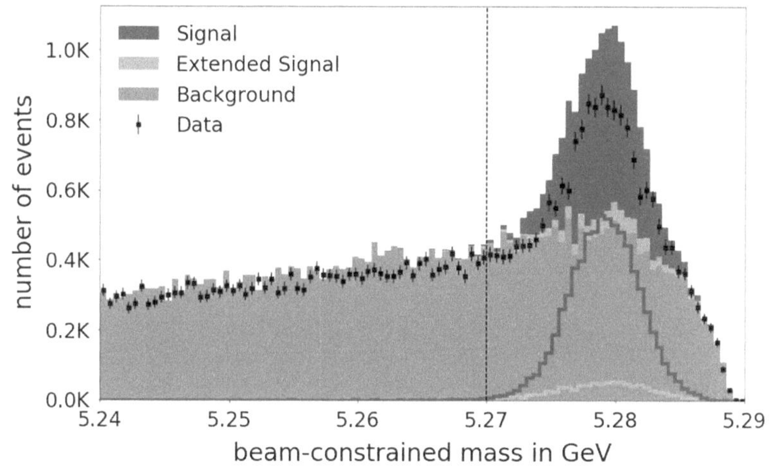

(b) A tight cut on the `SignalProbability` was performed.

Fig. 4.5 **Belle**: The beam-constrained mass of the best hadronically tagged B^+ meson candidate per event. The spectrum of the different signal-definitions is shown

4.3.2 Performance on Belle

The performance of the `FEI` was studied on converted Belle MC from the last Belle Monte Carlo campaign and converted Belle data recorded by the Belle detector. The

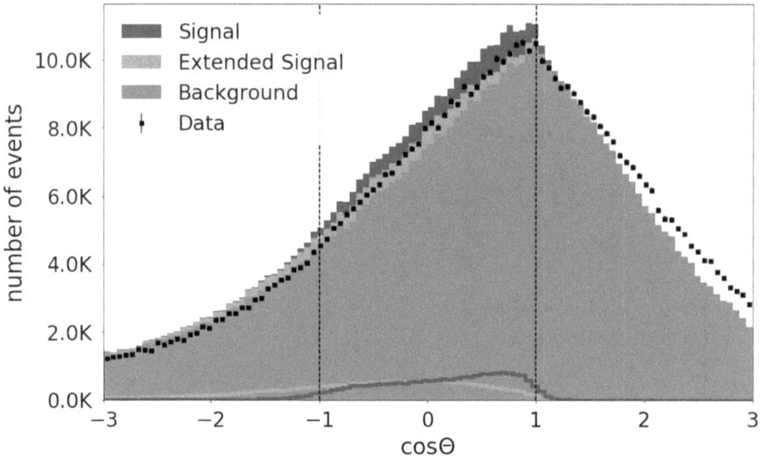

(a) A loose cut on the `SignalProbability` was performed.

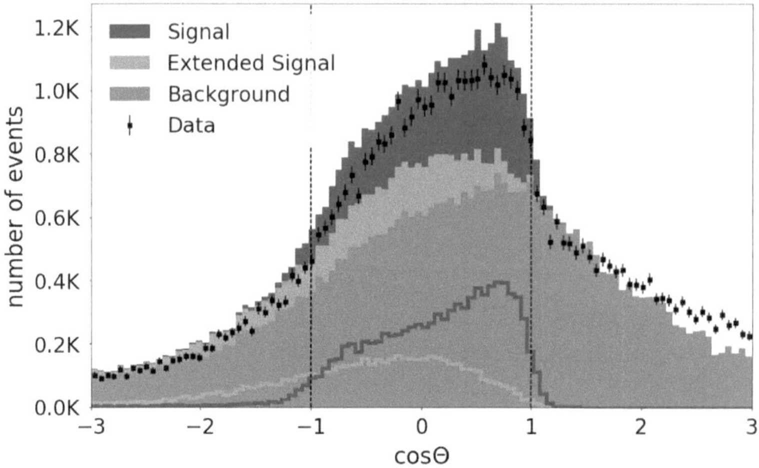

(b) A tight cut on the `SignalProbability` was performed.

Fig. 4.6 **Belle**: $\cos\Theta_{BD\ell}$ of the best semileptonically tagged B^+ meson candidate per event. The spectrum of the different signal-definitions is shown

sample used for the training of the `FEI` contained 200 million simulated $\Upsilon(4S)$ events divided evenly between the processes $\Upsilon(4S) \rightarrow B^+B^-$ and $\Upsilon(4S) \rightarrow B^0\bar{B}^0$. The sample used to estimate the performance of the `FEI` during the application phase contained 6 million simulated events divided evenly between the $\Upsilon(4S)$ processes $\Upsilon(4S) \rightarrow B^+B^-$, $\Upsilon(4S) \rightarrow B^0\bar{B}^0$, and 6 million simulated events divided

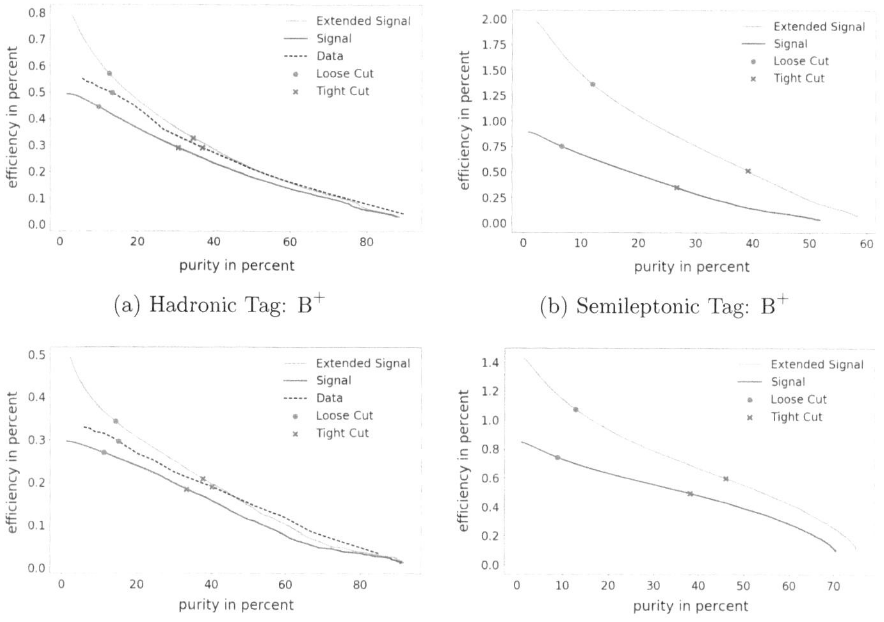

(a) Hadronic Tag: B^+

(b) Semileptonic Tag: B^+

(c) Hadronic Tag: B^0

(d) Semileptonic Tag: B^0

Fig. 4.7 Belle: Receiver operating characteristic (ROC) curve of tagged B mesons for the signal and extended signal-definition. The loose (`SignalProbability` > 0.01) and tight (`SignalProbability` > 0.1) cut are shown in orange and red, respectively

evenly between the continuum processes $e^+e^- \to c\bar{c}$, $s\bar{s}$, $d\bar{d}$ and $u\bar{u}$. In addition 10 million events recorded at the $\Upsilon(4S)$ resonance, and 10 million off-resonance events recorded 60 MeV below the $\Upsilon(4S)$ resonance were used. Throughout the text, the simulated events are scaled to the integrated luminosity of the recorded data they are compared to.

The generic `FEI` trained on all Belle events with less or equal to 14 tracks per event was applied. Table 4.3 shows the obtained tag-side efficiency and Table 4.4 shows the obtained tagging efficiency.

At this point an analysis with an exclusive signal-side would employ the completeness-constraint. Instead, only the candidate with the highest `SignalProbability` in each event is considered in the following. This best-candidate selection is independent of a specific signal-side, but not optimal (meaning the stated results can be understood as lower bounds for the tag-side efficiency and purity of the `FEI`).

4.3.2.1 Hadronic Tag-Side Efficiency Estimation on Data

The measurement of absolute branching fractions (in contrast to, e.g, branching fraction ratios) requires the knowledge of the selection efficiency on data. Therefore it is crucial to calibrate the tag-side efficiency of the `FEI` on data using control-channels (see Sect. 4.3.4). This section introduces an alternative method to estimate the tag-side efficiency, which does not require the reconstruction of a signal-side.

The hadronic tag-side efficiency can be estimated on recorded data by fitting the beam-constrained mass spectrum of the B meson candidates. The distribution of the signal candidates are modeled with a Gaussian distribution and the background candidates are modeled with an `ARGUS` function [1, Chap. 7.1]. The location and scale of the distributions were fixed to the Monte Carlo expectation. Hence, only the normalization and the shape parameter c of the `ARGUS` function are adjustable by the fit. The two components are fitted using an extended unbinned maximum likelihood (EUML) fit. Afterwards the tag-side efficiency and purity in a window of $5.24\,\text{GeV} < M_{bc} < 5.29\,\text{GeV}$ can be calculated using the fitted yields of the signal and background component.

This method overestimates the tag-side efficiency if the beam-constrained mass spectrum contains a background component that peaks at the same position as the signal component as can be seen in Fig. 4.8. Although the signal peak is clearly smaller on recorded data than on simulated events, the extracted signal fraction is the same, because the peaking background was assigned to the signal peak by the fit (see also the discussion of the signal-definition in Sect. 4.3.1).

The resulting efficiencies and purities for different cuts on the `SignalProbability` of the candidates is shown in Fig. 4.7. It is compatible with the expectation from Monte Carlo expectation, but suffers from the over-estimation problem discussed above.

A complementary and superior method using control-channels is discussed in Sect. 4.3.4. This method is robust against peaking background, because it does not rely on any signal-definition of the tag-side candidates.

The semileptonic tag-side efficiency can be estimated on recorded data by fitting the $\cos\Theta_{\mathrm{BD}\ell}$ spectrum of the B meson candidates. This was not further investigated during this thesis, but is described in more detail in [9].

4.3.2.2 Continuum Background

The performance of the `FEI` was studied on continuum data recorded by the Belle detector 60 MeV below the $\Upsilon(4S)$ resonance.

At this energy, the production of B mesons is kinematically forbidden and only non-resonant processes occur, hence the name "continuum data". Some kinematic properties like the beam-constrained mass are very sensitive to the reduced energy, therefore these properties are scaled to the correct energy scale. For instance, the beam-constrained mass is scaled by

$$S_{\text{energy}} = \frac{10.58\,\text{GeV}}{10.58\,\text{GeV} - 0.06\,\text{GeV}}.$$

to account for the shift of the kinematic end-point. In the case of $\cos\Theta_{BD\ell}$, the on-resonance beam-energy is used for the calculation (see Eq. 4.3).

As can be seen from Fig. 4.8 the continuum data does not contain any signal or peaking background, but the combinatorial background can be described well with an ARGUS function. The Monte Carlo simulation overestimates the amount of combinatorial background from continuum. The same holds true for the semileptonic B mesons (see Fig. 4.9), in addition the distribution is shifted. This is a known problem (already encountered in Sect. 3.3.3 and described in Sect. 2.3.3.3). The continuum background on data is about 20% lower than expected from Monte Carlo.

In total, the `FEI` performance does not seem to be influenced by the lower beam-energy of off-resonance data, which is a desired property because this allows the use of off-resonance data as a signal-free control region for continuum background.

4.3.3 Performance on Belle II

The performance of the `FEI` was studied on Belle II MC from 7th Belle II Monte Carlo campaign. The sample used for the training of the `FEI` contained 180 million simulated $\Upsilon(4S)$ events divided evenly between the processes $\Upsilon(4S) \rightarrow B^+B^-$ and $\Upsilon(4S) \rightarrow B^0\bar{B}^0$. The sample used to estimate the performance of the `FEI` during the application phase contained 7 million simulated events divided evenly between the processes $\Upsilon(4S) \rightarrow B^+B^-$, $\Upsilon(4S) \rightarrow B^0\bar{B}^0$, and $e^+e^- \rightarrow c\bar{c}$, $s\bar{s}$, $d\bar{d}$, $u\bar{u}$ and $\tau^+\tau^-$, and were scaled to an integrated luminosity of $1\,ab^{-1}$.

The generic `FEI` was trained on all Belle II events with less than or equal to 14 tracks per event. Table 4.5 shows the obtained tag-side efficiency. Table 4.6 shows the obtained tagging efficiency.

At this point an analysis with an exclusive signal-side would employ the completeness-constraint. Instead, only the candidate with the highest `SignalProbability` in each event is considered in the following. This cut is independent of a specific signal-side, but not optimal (meaning the stated results can be understood as lower bounds for the tag-side efficiency and purity of the `FEI`).

Table 4.5 Belle II: Maximum tag-side efficiency of the `FEI` on independent Monte Carlo simulated Belle II events

	Charged $B^\pm$	Neutral B^0
Hadronic (%)	0.66	0.38
Semileptonic (%)	1.45	1.94

Table 4.6 Belle II: Tagging efficiency of the `FEI` on independent Monte Carlo simulated Belle II events

	All		Signal region	
	Charged $B^\pm$	Neutral B^0	Charged $B^\pm$	Neutral B^0
Hadronic (%)	35.5	28.0	4.7	3.7
Semileptonic (%)	71.2	69.6	18.8	12.4

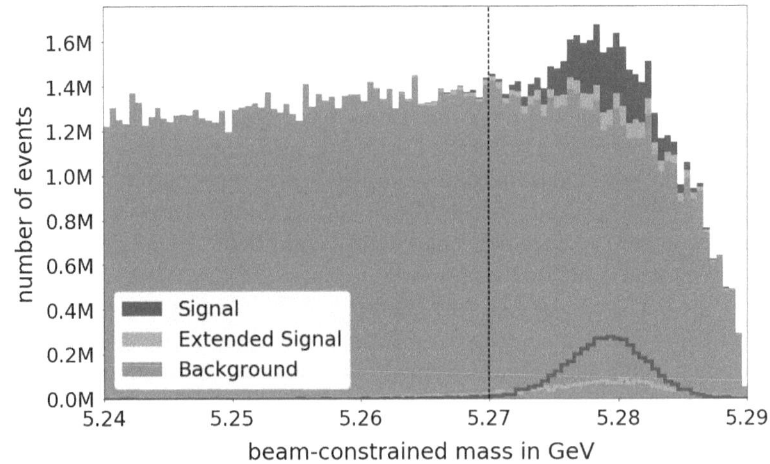

(a) A loose cut on the `SignalProbability` was performed.

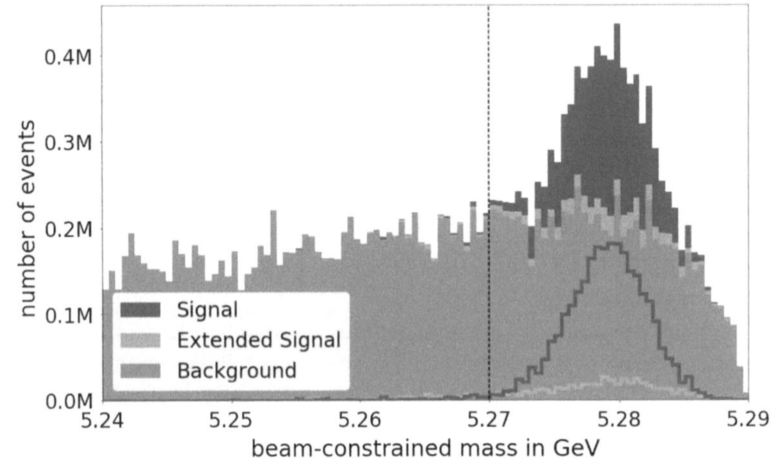

(b) A tight cut on the `SignalProbability` was performed.

Fig. 4.10 **Belle II**: The beam-constrained mass of the best hadronically tagged B^+ meson candidate per event The spectrum of the different signal-definitions is shown

Figure 4.10 shows the beam-constrained mass of the best hadronically tagged B^+ meson candidate per event. The correctly reconstructed B mesons peak clearly at the nominal B meson mass, as expected. Figure 4.11 shows $\cos \Theta_{BD\ell}$ of the best semileptonically tagged B^+ meson candidate per event. The correctly reconstructed B mesons peak clearly in the physically allowed region between -1 and 1, as expected. The dashed lines indicates the signal region which is used to calculate the tag-side efficiency and purity receiver operating characteristic curves in Fig. 4.12. The hadronic tag can reach a higher purity than the semileptonic tag, whereas the semileptonic tag provides a larger tag-side efficiency, as expected.

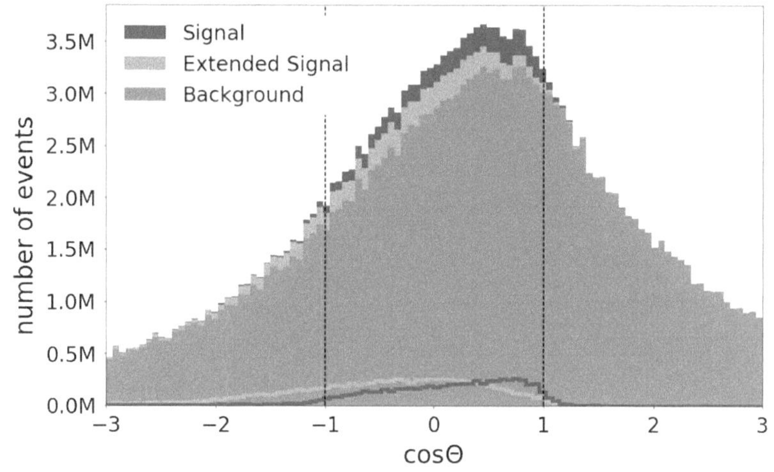

(a) A loose cut on the `SignalProbability` was performed.

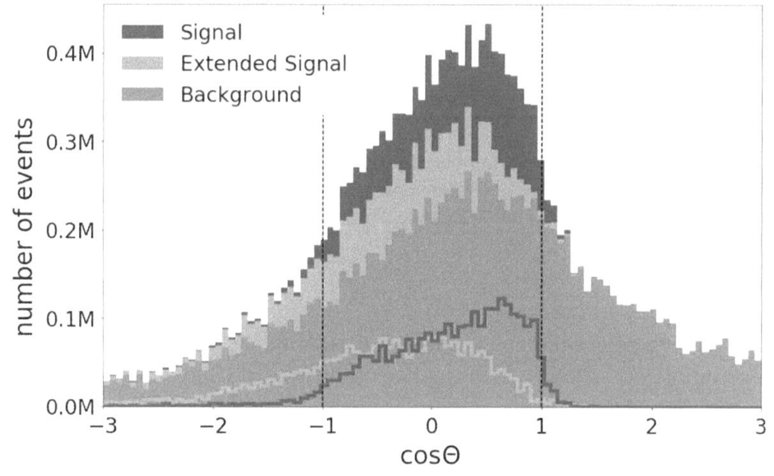

(b) A tight cut on the `SignalProbability` was performed.

Fig. 4.11 **Belle II**: $\cos\Theta_{BD\ell}$ of the best semileptonically tagged B^+ meson candidate per event. The spectrum of the different signal-definitions is shown

4.3.3.1 Influence of the Event Multiplicity

The number of reconstructed tracks (the **multiplicity**) is a fundamental property of an event. The larger the number of reconstructed tracks, the larger the combinatorial background and the more difficult it is to reconstruct the correct decay-chain. The influence of the event and candidate multiplicity on the performance of the `FEI` was investigated during this thesis.

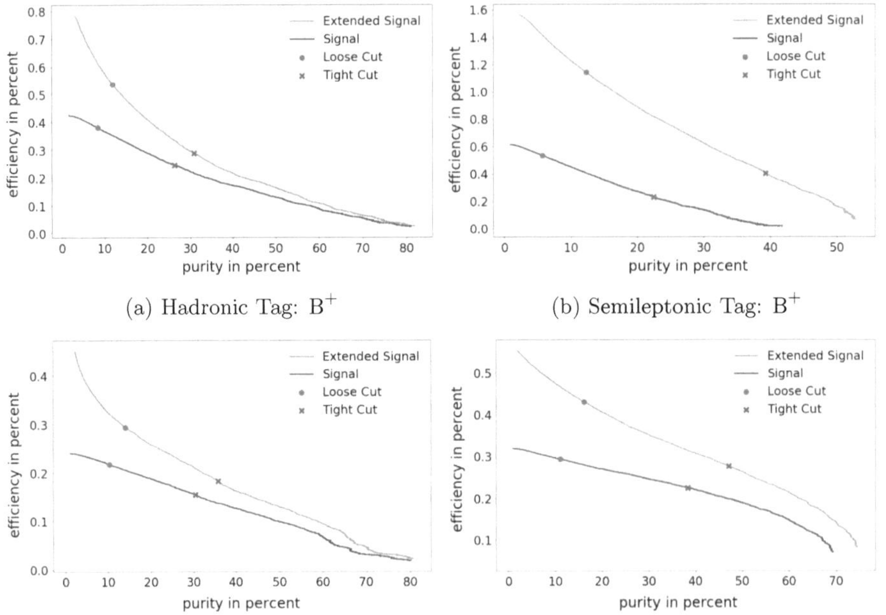

(a) Hadronic Tag: B^+ (b) Semileptonic Tag: B^+

(c) Hadronic Tag: B^0 (d) Semileptonic Tag: B^0

Fig. 4.12 **Belle II**: Receiver operating characteristic (ROC) curve of tagged B mesons for the signal and extended signal-definition. The loose (`SignalProbability` > 0.01) and tight (`SignalProbability` > 0.1) cut are shown in orange and red, respectively

Figure 4.13 shows the multiplicity of B_{tag} candidates. The majority of correctly reconstructed tag-side candidates consist of less than 8 tracks. Theoretically, the maximum number of tracks which can be combined to a valid charged (neutral) B_{tag} candidate by the `FEI` is 11 (12). However, in practice 9 (10) tracks per candidate is not exceeded.

As can be seen from Fig. 4.14 this effectively restricts the multiplicity of correctly tagged signal events. In this thesis the rare decay $B^+ \rightarrow \tau^+\nu_\tau$ is used as signal (see Chap. 5). Often the investigated rare B_{sig} decay-channel has a very low multiplicity N. Therefore it is reasonably to only consider events with less than $11 + N$ $(12 + N)$ reconstructed tracks in the training and/or in the application, because only those events will pass the completeness-constraint. Furthermore, it is expected that the `FEI` performs better on low-multiplicity events due to the reduced combinatorics, and that the required computing resources scales with the event multiplicity.

The influence of the maximum number of reconstructed tracks allowed in the training and in the application was studied. The study was conducted with version 3 of the `FEI`, whereas the rest of this thesis uses version 4. The results are shown in Fig. 4.15 and should be independent of the used version. Only the best B_{tag} candidate per event was considered, and an additional cut on the beam-constrained mass was applied.

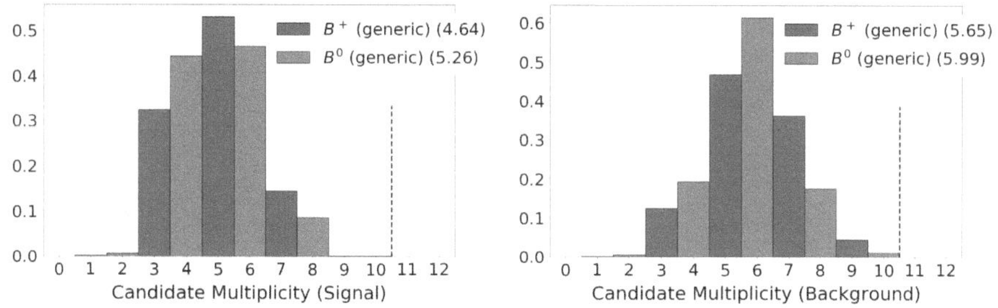

(a) The track multiplicity of correctly reconstructed B candidates.

(b) The track multiplicity of incorrectly reconstructed B candidates.

Fig. 4.13 Number of tracks (multiplicity) of the candidates. The dashed line indicates the default cut on the maximum numbers of tracks used during the application of the `FEI` in the benchmark analysis $B \to \tau\nu$ in this thesis. The number in parenthesis in the legend states the mean number of tracks for the corresponding component

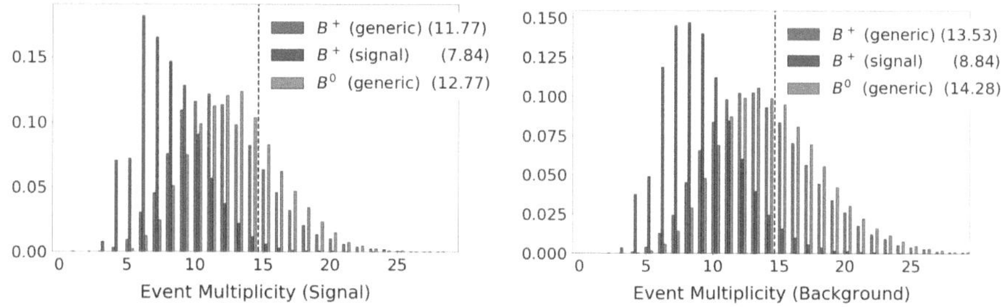

(a) The track multiplicity of the events with correctly reconstructed B candidates.

(b) The track multiplicity of the events with incorrectly reconstructed B candidates. An event is counted multiple times if it contains more than one background candidate.

Fig. 4.14 Number of tracks (multiplicity) of the events. The dashed line indicates the default cut on the maximum numbers of tracks used during the training of the `FEI` in this thesis. The rare decay $B^+ \to \tau^+\nu_\tau$ is used as signal. The number in parenthesis in the legend states the mean number of tracks for the corresponding component

Surprisingly, the maximum number of reconstructed tracks allowed in the training barely influences the performance of the `FEI` if applied to all events (see Fig. 4.15a). Hence, it is not important if the training is done on all events, or on low-multiplicity events. Only at a multiplicity as low as 6 a significant effect is visible. This can be explained by the hierarchical approach of the `FEI`. It allows the training of high-multiplicity B decay-channels using only low-multiplicity D channels. But during the application the high-multiplicity B decay-channels can also be combined with high-multiplicity D channels. In fact, a `FEI` trained on events with less than or equal to six tracks is still able to reconstruct tag-side candidates with a multiplicity of seven during the application quite well.

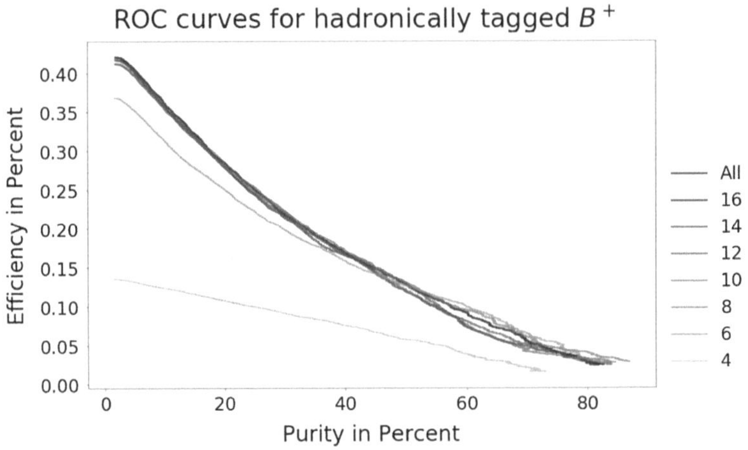

(a) Receiver operating characteristic (ROC) curves of the FEI applied to **all** events without a cut on the event multiplicity.

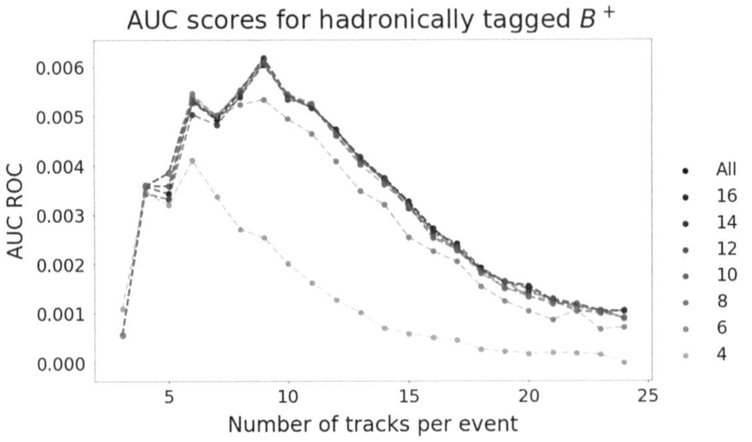

(b) The area under the ROC curve (AUC ROC) depending on the multiplicity of the event.

Fig. 4.15 Evaluation of the best hadronically tagged B^+ candidate per event, reconstructed by different configurations of the FEI. Each curve corresponds to a generic FEI with a cut on the maximum number of N tracks in the event during the **training**

Moreover, the performance of the FEI on events with different multiplicities is barely influenced by the maximum number of reconstructed tracks allowed during the training as well (see Fig. 4.15b). For instance, a FEI trained specifically on events with $\leq$6 tracks, does not perform better on events with 6 tracks than a FEI trained on all events. However, overall the FEI performs significantly better on low-multiplicity events, since it is much easier to reconstruct the correct B_{tag} meson. In particular, the tag-side efficiency for signal-sides with a low-multiplicity is usually significantly better (see Chap. 5).

The same behavior was found for hadronically tagged B^+, hadronically tagged B^0, semileptonically tagged B^+ and semileptonically tagged B^0.

The default training of the generic `FEI` considers only events with a maximum multiplicity of 14 during the training, and considers all events during the application. The number was chosen to save computing resources without any risk of decreasing the overall performance of the `FEI` on signal events (see Fig. 4.13). The benchmark analysis in Chap. 5 uses the default training, but applies an additional cut on the multiplicity during the application to save computing time without discarding signal events.

4.3.3.2 Influence of Beam-Background

The exact beam-background conditions at Belle II are currently (at the time of writing) unknown. The current estimations range between a 20 to 30 fold increase compared to Belle. The exact numbers depend on the detector region. Therefore the influence of the beam-background on the performance of the `FEI` was studied.

The generic `FEI` was **trained**, with (T1) and without (T0) beam-background. Both were **applied** to independent generic $\Upsilon(4S)$ events, with (A1) and without (A0) beam-background. The results are shown in Fig. 4.16. Only the best B_{tag} candidate per event was considered, and an additional cut on the beam-constrained mass was applied.

As expected, T0 performs better than T1 on A0, and T1 performs better than T0 on A1.

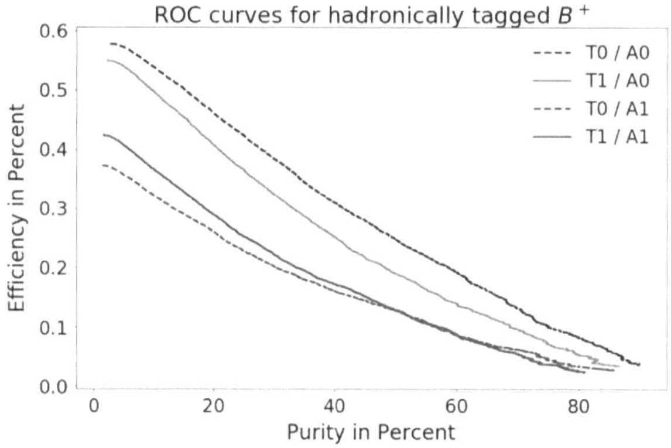

Fig. 4.16 Evaluation of the best hadronically tagged B^+ candidate per event, reconstructed by different configurations of the `FEI`. The solid (dashed) curves correspond to a generic `FEI` trained on simulated Belle II events with (without) beam-background. The blue and purple (red and orange) curves correspond to a generic `FEI` applied to simulated Belle II events with (without) beam-background

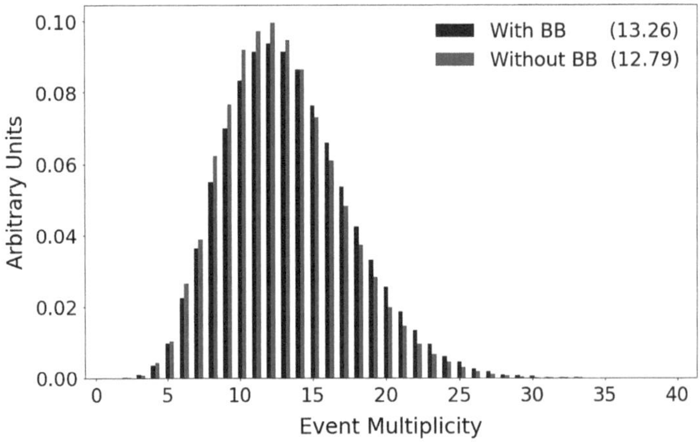

Fig. 4.17 The distribution of the event multiplicity for $\Upsilon(4S)$ events with and without beam-background (BB). On average, there is a fake track due to beam-background in every second event. The number in parenthesis in the legend states the mean number of tracks for the corresponding component

The training of the `FEI` is robust against beam-background. The loss in tag-side efficiency due to occurring beam-background in the training (that is the difference between T1 and T0) is of the order of 10%.

In contrast, the application of the `FEI` is sensitive to the beam-background. The loss in tag-side efficiency due to occurring beam-background in the application (that is the difference between A1 and A0) is of the order of 30%.

In particular the increased fake-rate (the fraction of reconstructed tracks which do not originate from a charged particle) caused by the increased beam-background is problematic (see Fig. 4.17), because it leads to increased combinatorics and decreases the probability of identifying the correct B_{tag} candidate.

The same behavior was found for hadronically tagged B^+, hadronically tagged B^0, semileptonically tagged B^+ and semileptonically tagged B^0.

Finally, the completeness-constraint is very sensitive to an increased fake-rate. Also quantities like the extra energy in the electromagnetic calorimeter are sensitive to the amount of beam-background. However, it is expected that the fake-rate will be significantly reduced in the upcoming years, because the reconstruction software of Belle II is not yet optimized to reduce the fake-rate. This is further discussed in the benchmark analysis in Chap. 5.

4.3.3.3 Influence of Reduced Boost

The asymmetry between the electron and positron beam of SuperKEKB was reduced with respect to KEKB, as stated in Table 2.2. The associated change of the Lorentz boost reduces the resolution of the decay time difference in time-dependent CP

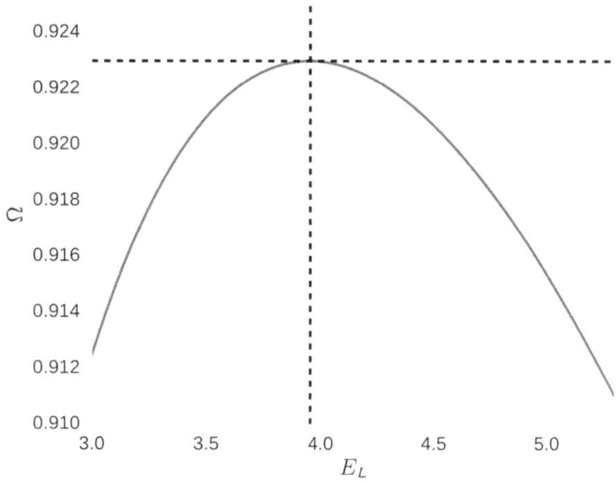

Fig. 4.18 Geometrical detector acceptance in a boosted frame of reference depending on the energy E_{L} of the positron beam

violation measurements, on the other hand it increases the detector acceptance. In spherical coordinates the Belle (and Belle II) detector covers $\Theta \in [17°, 150°]$ and $\phi \in [0°, 360°]$. The geometric detector acceptance in percent is

$$\Omega = \frac{1}{4\pi} \int_{17°}^{150°} \int_{0°}^{360°} \mathrm{d}\Omega = 91.1\%.$$

Assuming a uniform spherical distribution of the tracks and neglecting the mass of the final state particles compared to their momenta, the angle $\Theta^{\star}$ in the boosted frame of reference is

$$\tan\left(\Theta^{\star}\right) = \frac{1}{\gamma}\left(\frac{\sin\Theta}{\beta + \cos\Theta}\right)$$

where $\gamma = \frac{E_{\mathrm{H}} - E_{\mathrm{L}}}{E_{\mathrm{CMS}}}$ and $\beta = \frac{E_{\mathrm{H}} - E_{\mathrm{L}}}{E_{\mathrm{H}} + E_{\mathrm{L}}}$. In consequence the geometrical detector acceptance in the boosted frame of reference depends on the Lorentz boost of the collider. As can be seen from Fig. 4.18 this effect is of the order of sub-percent. The `FEI` requires the reconstruction of roughly five to seven tracks in the event, consequently the reduced boost does have a significant influence on the sensitivity [10].

4.3.4 Tag-Side Efficiency Correction

The uncertainty on the tag-side efficiency of the `FEI` is (one of) the most important systematic uncertainties in the measurement of branching fractions of rare decays, e.g. $\mathcal{BR}(\mathrm{B}^{+} \to \tau^{+}\nu_{\tau}) = (1.25 \pm 0.28 \pm 0.27) \times 10^{-4}$ [11], with a systematic uncertainty due to the tag-side efficiency of 0.16×10^{-4}.

Therefore it is indispensable to calibrate and correct the tag-side efficiency on data using multiple control channels.

The FEI was calibrated on converted Belle data within the scope of a supervised master's thesis [12]. I briefly summarize the results here.

The FEI was applied to the full $\Upsilon(4S)$ dataset recorded by Belle. The well-known branching fractions of ten control channels were measured using the FEI. The difference between the Monte Carlo expectation and the data yields tag-side channel dependent calibration factors for the tag-side efficiency, as well as their uncertainties.

Figure 4.19 summarizes the results for the ten control channels. The overall calibration factors for the latest FEI version 4 averaged over all control channels are

$$\varepsilon_{\text{charged}} = 0.74^{+0.014}_{-0.013} \pm 0.050$$
$$\varepsilon_{\text{mixed}} = 0.86^{+0.045}_{-0.050} \pm 0.054.$$

The measured values are compatible with the observed lower tag-side efficiency on recorded data in comparison with the Monte Carlo expectation in Fig. 4.8.

A similar study was conducted in the past for the FR by [13], yielding a similar overall factor of $\varepsilon = 0.75 \pm 0.03$. The rather large discrepancy between simulated

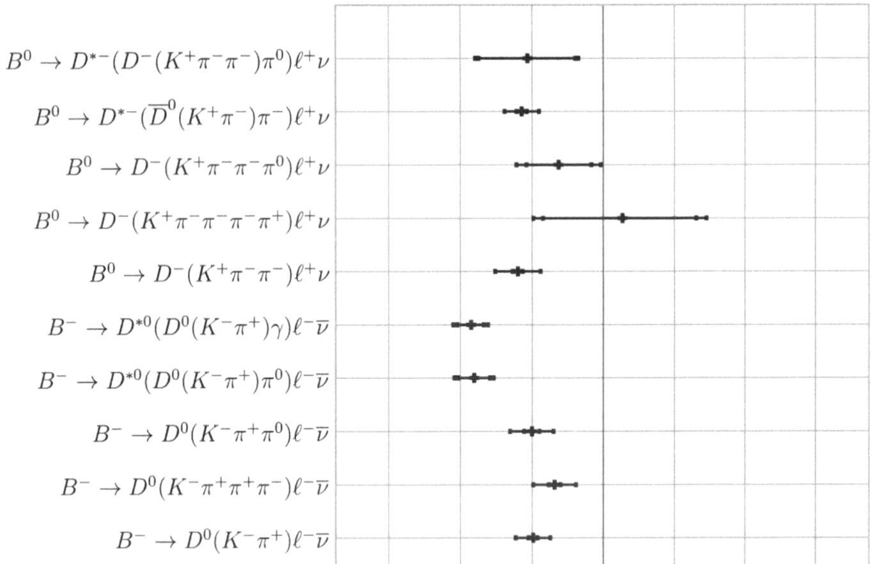

Fig. 4.19 The overall efficiency correction calculated by measuring the known branching fractions of 10 control channels on converted Belle data [12]. The calibration was performed on the latest FEI version 4

events and recorded data is caused by the uncertainty on the branching fractions and decay models of the simulated B decay-channels used for the tag-side and the large number of multivariate classifiers involved in the process.

The tag-side-efficiency can be corrected using the measured calibration factors. A measurement which uses these factors to correct the tag-side-efficiency is performed relative to the considered calibration decay channels. The systematic uncertainty of the correction is given by the uncertainty of the factors.

The uncertainty of the tag-side efficiency has only a minor effect on searches of rare decays, which are statistically limited, e.g., in the search for $B \to K\nu\bar{\nu}$ (see [14]) or $B \to \ell\nu\gamma$ (see [15]). In addition, the uncertainty on the tag-side efficiency can be removed by performing a relative measurement as the efficiency will cancel out due to the normalization, e.g., $R = \frac{\mathcal{BR}(B \to D\tau\nu_\tau)}{\mathcal{BR}(B \to D\ell\nu)}$ (see [16]).

4.3.5 Comparison with Full Reconstruction

The performance of the FEI was compared to the FR on converted Belle MC from the last Belle Monte Carlo campaign. All numbers shown here refer to the Monte Carlo expectation. As was shown in Sect. 4.3.4, the performance on data for the FEI is about 20% lower.

The tag-side efficiencies of the FR are stated in Table 4.7. The tag-side efficiencies of the FEI using an identical configuration, meaning the same decay-channels, are stated in Table 4.8.

Using the identical configuration, the FEI outperforms the FR, and increases the hadronic tag-side efficiency by 89% (83%) for charged (neutral) B mesons. This increase in tag-side efficiency is due to the best-candidate selection and improved classifiers. Using the default configuration, the hadronic tag-side efficiency is increased by 171% (156%) for charged (neutral) B mesons. This further increase in tag-side efficiency is due to the additional new decay-channels. The stated results for the hadronic tag were verified on data (see Sects. 4.3.2.1 and 4.3.4).

The official FR did not provide a semileptonic tag. However, variants of the FR were used for semileptonic tagging (see [9, 11]). The FEI significantly improves the semileptonic tag-side efficiency, however the stated results are based on Monte Carlo simulation and are not yet validated on data.

Table 4.7 Tag-side efficiency of the FR [9]

	Charged $B^\pm$	Neutral B^0
Hadronic (%)	0.28	0.18
Semileptonic (%)	0.31	0.34

Table 4.8 Tag-side efficiency of the FEI using the identical configuration as the FR

	Charged $B^\pm$	Neutral B^0
Hadronic (%)	0.53	0.33

The performance of the FEI on Belle II MC events is worse than on Belle events, but still significantly better than the FR on Belle events. The hadronic tag-side efficiency is increased by 135% (111%) for charged (neutral) B mesons. As shown in Sect. 4.3.3.2 the FEI performance is sensitive to the fake-rate due to the increased beam-background. It is expected that the fake-rate will be drastically reduced as soon as the Belle II reconstruction software is finished and optimized.

In consequence, the stated increases in tag-side efficiency for Belle II are a lower bound of the expected improvements.

Altogether, the improvements of the FEI with respect to the FR are very consistent and explainable for all combinations of tag (hadronic or semileptonic), charge (charged or neutral), experiment (Belle or Belle II) and data type (MC events or recorded data).

4.4 Outlook

The design of the FEI foresees the possibilities to extend the algorithms in the future. In this section, I summarize ideas and techniques that were not further investigated during this thesis. However, they have a large potential to further increase the key performance indicators (tagging efficiency, tag-side efficiency and purity) of the FEI.

4.4.1 *Inclusive Tagging Using the FEI*

The FEI can be used for of inclusive tagging. Preliminary studies on the signal channel $B^+ \to \mu^+ \nu_\mu$ have shown that this is possible in principle. The FEI was applied to the rest-of-event of a reconstructed B_{sig}, and the intermediate pre- and post-cuts were relaxed as far as possible. A valid candidate was then created in $\mathcal{O}(10\%)$ of the cases.

The resulting four-momentum and vertex position is as good as that provided by the traditional inclusive tag. Now the suspected explicit decay-chain can be used to further improve the four-momentum and vertex position with the help of a decay tree fitter [4].

BASF2 does not include a usable decay tree fitter at the time of writing. Hence, further studies have been postponed.

4.4.2 $D \to X \ell \nu$

Semileptonic tagging provides a higher efficiency compared to its hadronic counterpart, but due to the missing kinematic information the tagged sample is not as pure. Until now, only semileptonic B decays were used for this.

However, semileptonic D decays can be used as well and behave similarly. 13% of the D^0 and 33% of the D^+ mesons decay semileptonically. The B meson is reconstructed in a hadronic decay-channel $B \rightarrow D(\ldots)$, and the D meson is reconstructed in a semileptonic decay-channel $D \rightarrow (\ldots)\ell\nu$. Using the known invariant mass of the D and the possibly occurring D^* meson, the missing kinematic information due to the neutrino can be partially recovered.

The current default `FEI` configuration (see Sect. B.1) already includes in addition semileptonic D decays.

Although, this new class of decay-channels yields impure tagged samples, it provides the next step towards even more inclusiveness, and can be beneficial in situations where the `FEI` would otherwise not output a valid candidate at all.

4.4.3 $X \rightarrow YK_L^0$

A generic B decay contains in 29% of the cases a K_L^0. A K_L^0 can be detected by the Belle and the Belle II detector, but both provide only a rough energy and flight direction estimation. The situation can be compared to semileptonic decay-channels, where kinematic information is missing as well. Until now, decay-channels containing K_L^0 particles were not used.

Some Belle analyses employed a K_L^0 veto. In consequence, tagged events with a detected K_L^0 were rejected. This situation can be improved, especially in view of Belle II, which will provide a better K_L^0 identification.

The current default `FEI` configuration (see Sect. B.1) already includes in addition decay-channels with a K_L^0.

Although, this new class of decay-channels yields impure tagged samples, it provides the next step towards even more inclusiveness, and can be beneficial in situations where the `FEI` would otherwise not output a valid candidate at all.

4.4.4 $\pi^0 \rightarrow \gamma(\gamma)$

Many B and D decay-channels include π^0 particles, which decay immediately into two photons. A generic B decay contains in 89% of the cases a π^0.

Assuming a spherical distribution and a geometric detector acceptance of 91.1% (by taking the boost into consideration as shown in Sect. 4.3.3.3) 14% of the π^0 will have one undetectable photon due to the acceptance. This number was confirmed using MC simulation.

The `FEI` could use π^0 candidates reconstructed from just one photon, and check if an assumed second photon could be outside of the detector acceptance. Using the known invariant mass of the π^0 and the detector acceptance, the momentum of the missing photon could be estimated, leading eventually to the recovery of the incomplete π^0 particles.

As in the case of semileptonic decay-channels and decay-channels with K_L^0, the FEI could profit even from incomplete π^0 particles in situations where the FEI would otherwise not output a valid candidate at all. Furthermore, many measurements of rare decays, which use exclusive tagging, extract the signal-yield from the extra energy in the electromagnetic calorimeter E_{ECL} (see Chap. 5). If the FEI assigns all tracks correctly to a tag-side B_{tag} meson, but misses an incomplete π^0, one additional photon from the tag-side will end up in E_{ECL}. Therefore, reconstructing incomplete π^0 has the potential to improve the resolution of the signal peak in E_{ECL}.

4.4.5 FEI on $\Upsilon(5S)$

The FEI can directly be applied to the $\Upsilon(5S)$ resonance. This resonance decays into a pair of BB(5.5%), BB*(13.7%), B*B*(38.1%), $B_s^*B_s^*$(20.1%) mesons. The user has to add the B* and B_s^* particles and their associated decay-channels to the FEI configuration. The powerful completeness-constraint can still be applied in this situation.

4.4.6 Deep FEI

Newly developed algorithms based on deep-learning can outperform established algorithms, as was shown in the case of flavour tagging (Sect. 3.3.1.2) and continuum suppression (Sect. 3.3.1.3). Both algorithms can be seen as a form of inclusive tagging, hence, it is obvious to study the applicability of deep-learning to exclusive tagging as well [19].

As described in Sect. 3.1.4.3, applications with a learnable distributed representation, are in particular suitable for deep-learning. The most successful applications of deep-learning use domain-specific knowledge to restrict the learnable representation, e.g. by using convolutional neural networks in image recognition (see [17, Chap. 9]), recurrent neural network in speech recognition (see [17, Chap. 10]), and relational neural networks for relational reasoning (see [18]).

In the case of exclusive tagging a reasonable representation of the input data (tracks and clusters) is already known from physics: it is the decay-chain of the $\Upsilon(4S)$. This representation has all desired properties defined in Sect. 3.1.4.3. The FEI is designed to reconstruct this decay-chain in an hierarchical approach: starting from the low-level information (tracks and cluster), creating intermediate particle candidates and finally providing the most likely B meson candidates.

In that sense the FEI is a deep-learning based algorithm, which enforces the decay-chain representation. It comes as no surprise that the hierarchical approach can be visualized by a network (see Fig. 4.3), where the nodes are given by the particle and the weights are multivariate classifiers. In contrast to an ordinary deep neural network, the `FEI` can take advantage of complex existing algorithms in each step, e.g. vertex-fitting. Furthermore, the explicit decay-chain is automatically known, whereas it is not straight-forward to design a deep neural network that outputs the explicit decay-chain it reconstructed.

4.5 Conclusion

The `Full Event Interpretation` enables Belle II physicists to measure a wide range of interesting decays with a minimum of detectable information. It exploits the unique experimental setup of B factories.

During this thesis the `FEI` was further developed from a "Proof of Concept" to a standard tool used in production. The algorithm was extensively studied on Belle and Belle II MC, and the obtained results were verified on data recorded by the Belle detector using the `b2bii` package.

The `FEI` more than doubles the tag-side efficiency for all combinations of tag (hadronic/semileptonic), charge (charged/neutral) and experiment (Belle/Belle II), compared to its (already very successful) predecessor. The tag-side efficiency for hadronically tagged B mesons was calibrated on Belle data, and is ready for the productive usage in analyses on converted Belle data.

Besides, many software tools originally developed for the `FEI` like `FastBDT` and `FastFit` proofed useful in many other contexts. The `FEI` was the first algorithm to take full advantage of the `b2bii` (see Chap. 2) and `mva` (see Chap. 3) package.

Finally, there are many possible extensions and unexplored applications of the `FEI`, which provide an exciting and fruitful area for further research.

References

1. A.J. Bevan et al., The physics of the B factories. Eur. Phys. J. C **74**, 3026 (2014). https://doi.org/10.1140/epjc/s10052-014-3026-9
2. T. Keck, The full event interpretation for Belle II. M.A. thesis. KIT, 2014, https://ekp-invenio.physik.uni-karlsruhe.de/record/48602
3. M. Feindt et al., A hierarchical neurobayes-based algorithm for full reconstruction of B mesons at B factories. Nucl. Instrum. Methods **A654**, 432–440 (2011). https://doi.org/10.1016/j.nima.2011.06.008
4. W.D. Hulsbergen, Decay chain fitting with a Kalman filter. Nucl. Instrum. Methods **A552**, 566–575 (2005). https://doi.org/10.1016/j.nima.2005.06.078
5. W. Waltenberger, W. Mitaroff, F. Moser, RAVE—a Detector-independent vertex reconstruction toolkit. Nucl. Instrum. Methods Phys. Res. Sect. A: Accel. Spectrom. Detect. Assoc. Equip. **581**(1–2), 549–552 (2007), https://doi.org/10.1016/j.nima.2007.08.048
6. R. Fruhwirth, Application of Kalman filtering to track and vertex fitting. Nucl. Instrum. Methods (1987). https://doi.org/10.1016/0168-9002(87)90887-4

7. G. M. Amdahl, Validity of the single processor approach to achieving large scale computing capabilities, in *Proceedings of the Spring Joint Computer Conference*, pp. 483–485, 18–20 April 1967, https://doi.org/10.1145/1465482.1465560
8. https://github.com/thomaskeck/FastFit. Accessed 02 Oct 2017
9. K. Kirchgessner, Semileptonic tag side reconstruction. M.A. thesis. KIT, 2012, http://ekp-invenio.physik.uni-karlsruhe.de/record/48181
10. I. Adachi et al., Belle design study report (2008), arXiv: 0810.4084 [hep-ex]
11. B. Kronenbitter et al., Measurement of the branching fraction of $B^+ \to \tau^+\nu_\tau$ decays with the semileptonic tagging method. Phys. Rev. D **92**(5), 051102 (2015). https://doi.org/10.1103/PhysRevD.92.051102
12. J. Schwab, Calibration of the full event interpretation for the Belle and the Belle II experiment. M.A. thesis, KIT, 2017, https://ekp-invenio.physik.uni-karlsruhe.de/record/48931
13. A. Sibidanov et al., Study of exclusive $B \to X_u l\nu$ decays and extraction of $|V_{ub}|$ using full reconstruction tagging at the Belle experiment. Phys. Rev. D (2013). https://doi.org/10.1103/PhysRevD.88.032005
14. J. Grygier, Search for $B \to h\nu\nu$ decays with semileptonic tagging at Belle (2017), arXiv: 1702.03224 [hep-ex]
15. A. Heller, Search for $B^+ \to \ell^+\nu\gamma$ decays with hadronic tagging using the full Belle data sample. Ph.D. thesis. KIT, 2015, https://ekp-invenio.physik.unikarlsruhe.de/record/48743
16. M. Huschle, Measurement of the branching ratio of $B \to D^{(*)}\tau\nu_\tau$ relative to $B \to D^{(*)}\ell\nu_\ell$ decays with hadronic tagging at Belle. Ph.D. thesis. KIT, 2015, https://ekp-invenio.physik.uni-karlsruhe.de/record/48622
17. Y. Lecun, Y. Bengio, G. Hinton, Deep learning. Nature **521**, 436–444 (2015). https://doi.org/10.1038/nature14539
18. A. Santoro et al., A simple neural network module for relational reasoning (2017), arXiv: 1706.01427 [cs.CL]
19. J. Schmidhuber, Deep learning in neural networks: An overview. Neural Netw. (2015). https://doi.org/10.1016/j.neunet.2014.09.003

Chapter 5
$B \rightarrow \tau\nu$

The leptonic decays of charged B mesons belong to the golden decay-channels at B factories. The expected branching fractions in the Standard Model of particle physics (SM) are small, hence leptonic decays are so-called **rare-decays**. The theoretical uncertainties on the predicted branching fraction are small and the possible influence of physics beyond the Standard Model (BSM) can be large.

The experimental signature is (mostly) a single track produced by a lepton or its decay product. The accompanying neutrinos are not detected. The leptonic decays are currently only accessible via the unique experimental setup at B factories.

The tauonic decay $B \rightarrow \tau\nu$, is of particular interest. Its branching fraction is the largest among the leptonic decays, due to the high mass of the τ meson elevating the helicity suppression. On the other hand the τ is unstable in the detector, hence it can only be measured indirectly using its decay products. Multiple neutrinos in the final state render this decay particularly experimentally challenging. It is the most prominent use-case for exclusive tagging.

During this thesis the decay $B \rightarrow \tau\nu$ was investigated as a benchmark analysis for the developed software tools and algorithms presented in the preceding chapters. The previous Belle analyses [1, 2] were repeated on the full $\Upsilon(4S)$ dataset recorded by Belle using the `b2bii` package. Furthermore, a sensitivity study for Belle II was conducted.

The main goal was to validate the developed tools and establish a prototype analysis, which can be used as a starting point for other analyses based on `b2bii` and/or the `FEI`.

In this chapter, I describe the theoretical background (Sect. 5.1), the experimental measurement (Sect. 5.2), the validation of the analysis strategy using off-resonance data and sidebands (Sect. 5.3), and the obtained results (Sect. 5.4).

T. Keck, *Machine Learning at the Belle II Experiment*, Springer Theses,
https://doi.org/10.1007/978-3-319-98249-6_5

5.1 Theory

The Standard Model of particle physics (SM) describes the elementary particles and their fundamental interactions, except for gravity.

The SM is a relativistic quantum field theory, that is the particles are described by quantum fields, whose equations of motion are invariant under the Poincaré space-time symmetry. The fundamental interactions (the electromagnetic, strong and weak interaction) are described by local gauge symmetries. The field equations of the SM are encoded in a Lagrangian density $\mathcal{L}$, from which the field equations can be derived using the Lagrange formalism.

The SM (with massless left-handed neutrinos) has 19 free parameters. All of them can and have been experimentally determined. A thorough introduction to the SM can be found in [3].

5.1.1 *Leptonic* $B^{\pm}$ *Meson Decays in the Standard Model*

The branching fraction of the leptonic decay of charged $B^{\pm}$ mesons can be calculated in the SM using perturbation theory. The leading order Feynman Diagram is shown in Fig. 5.1. It results in the following expression for the branching fraction [4, Sect. 7.10.2]

$$\mathcal{B}\left(B^{+} \to \ell^{+}\nu\right)_{SM} = \frac{G_F^2 M_B M_\ell^2}{8\pi}\left(1 - \frac{M_\ell^2}{M_B^2}\right)^2 f_B{}^2 |V_{ub}|^2 \tau_B, \tag{5.1}$$

where all occurring constants can be theoretically calculated and/or independently experimentally measured. The numerical values and their relative uncertainties are summarized in Table 5.1. The measurement of the branching fraction of the leptonic decay of charged B mesons can reduce the existing uncertainty, and provides an independent verification for the consistency of the SM with the experimental observations.

The leading dependency of the branching fraction on the squared lepton mass M_ℓ^2 is caused by the so-called helicity suppression [5, p. 1109]. In consequence, the τ

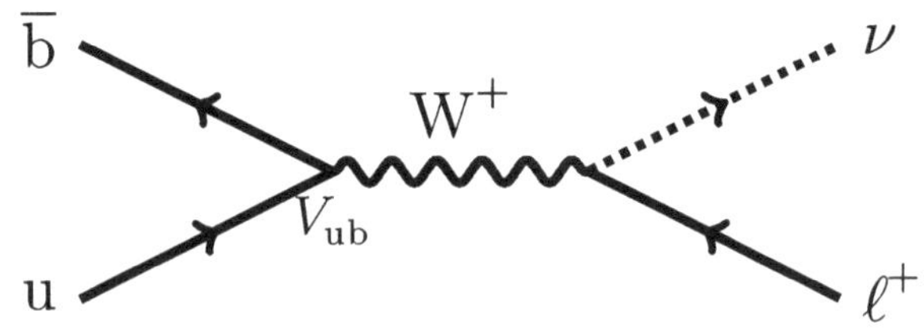

Fig. 5.1 Feynman Diagram of the leading standard model process mediated by a charged W boson for the purely leptonic decay of a charged B meson

Table 5.1 Numerical values and relative uncertainties of all quantities required to calculate the branching fraction of $B^+ \to \ell^+\nu$ in the standard model. All values are taken from the PDG review 2016 [5]

	PDG value	Relative uncertainty
G_F	$11.7\,\mathrm{TeV}^{-2}$	$5 \cdot 10^{-7}$
m_B	$5.28\,\mathrm{GeV}$	$3 \cdot 10^{-5}$
m_τ	$1.78\,\mathrm{GeV}$	$7 \cdot 10^{-5}$
τ_B	$1.64\,\mathrm{ps}$	$2 \cdot 10^{-3}$
f_B	$187.1\,\mathrm{MeV}$	$2 \cdot 10^{-2}$
$\lvert V_{ub}\rvert_{inc}$	$4.49 \cdot 10^{-3}$	$5 \cdot 10^{-2}$
$\lvert V_{ub}\rvert_{exc}$	$3.72 \cdot 10^{-3}$	$5 \cdot 10^{-2}$
$\lvert V_{ub}\rvert_{avg}$	$4.09 \cdot 10^{-3}$	$1 \cdot 10^{-1}$

Table 5.2 Standard model prediction for $B^+ \to \ell^+\nu$ calculated by Eq. 5.1 using the numerical values given in Table 5.1 and the averaged $|V_{ub}|$ value

	SM prediction	PDG 2016
$\mathcal{B}(B^+ \to e^+\nu_e)$	$(1.09 \pm 0.21) \cdot 10^{-11}$	$<9.8 \cdot 10^{-7}$ CL = 90%
$\mathcal{B}(B^+ \to \mu^+\nu_\mu)$	$(4.65 \pm 0.91) \cdot 10^{-7}$	$<1.0 \cdot 10^{-6}$ CL = 90%
$\mathcal{B}(B^+ \to \tau^+\nu_\tau)$	$(1.03 \pm 0.2) \cdot 10^{-4}$	$(1.06 \pm 0.20) \cdot 10^{-4}$

lepton is expected to have the largest branching fraction among the leptonic B decay channels, as can be seen in Table 5.2.

5.1.1.1 The V_{ub} Puzzle

The CKM matrix element $|V_{ub}|$ corresponds to one of the 19 free parameters of the SM. It induces the dominant theoretical uncertainty in the prediction of the branching fraction of $B^+ \to \ell^+\nu$. $|V_{ub}|$ can be determined by several distinct measurements, which differ in their associated theoretical and experimental uncertainties.

- The **full decay width of the inclusive decay** $B \to X_u\ell\nu$ can be calculated with high precision, that is <5% theoretical uncertainty, since QCD form-factors are not required. The inclusive decay has a high branching fraction, but the experimental measurement is difficult due to the high background contamination from the CKM-favored $B \to X_c\ell\nu$ decay.
- Therefore it is easier to measure **the partial decay width of the inclusive decay** $B \to X_u\ell\nu$, by demanding a high momentum for the lepton. However, the theoretical prediction of a partial decay width in this phase-space region is much more difficult and requires the knowledge of a non-perturbative distribution function, the so called "shape-functions" [5, p. 1317].
- The **exclusive measurement of the branching fraction** of $B \to \pi\ell\nu$ is experimentally clean, but has a lower branching fraction and requires in addition form-factors on the theoretical side. The measurement can be performed with and without exclusive tagging. Both approaches yield compatible results [5, p. 1319].

There is a slight tension between the determination of $|V_{ub}|$ from inclusive and exclusive measurements, as can be seen in Table 5.1. Patrignani et al. [5] states that the average value of $|V_{ub}|$ should be treated with caution "*given the poor consistency between the two determinations*". Nevertheless, this thesis uses the average value (see Table 5.1), but notes the differences between the $|V_{ub}|$ from exclusive and inclusive measurements where appropriate.

Finally, the measurement presented in this thesis can provide an independent determination of $|V_{ub}|$. The theoretical prediction of the branching fraction of B $\rightarrow \ell\nu$ is theoretically easy, since there are no strongly interacting particles in the final state. Experimentally it is challenging since there is usually only one measurable track.

5.1.1.2 The f_B Decay Constant

The contribution of the strong interaction in the decay B $\rightarrow \ell\nu$ is represented by the decay constant f_B. It is related to the overlap of the bottom and up quark wavefunctions in the meson.

The decay constant induces the second largest theoretical uncertainty besides the CKM matrix element $|V_{ub}|$. It can be calculated using lattice QCD simulations and QCD sum rules. Both approaches yield compatible results. This thesis uses the average value (see Table 5.1) from lattice QCD simulations determined by [5, p. 1117].

5.1.2 Type II Two Higgs Doublet Model

New physics contributions could influence the measurement of the branching fraction of B $\rightarrow \ell\nu$. The most famous model considered in the literature in connection with B $\rightarrow \ell\nu$ is the Type II Two Higgs Doublet Model [6]. In this model, the Standard Model Higgs Doublet is replaced with two Higgs Doublets, where one couples to down quarks and charged leptons, while the other couples to up quarks. The model introduces additional parameters: tan β, that is the ratio of the two vacuum expectation values of the two Higgs Doublets, and the masses $M_{H^\pm}$, M_{H^0} and M_{A^0} of the four additional Higgs particles.

The model provides another leading order Feynman Diagram shown in Fig. 5.2, in which the $W^\pm$ boson is replaced by a hypothetical charged Higgs. The two diagrams

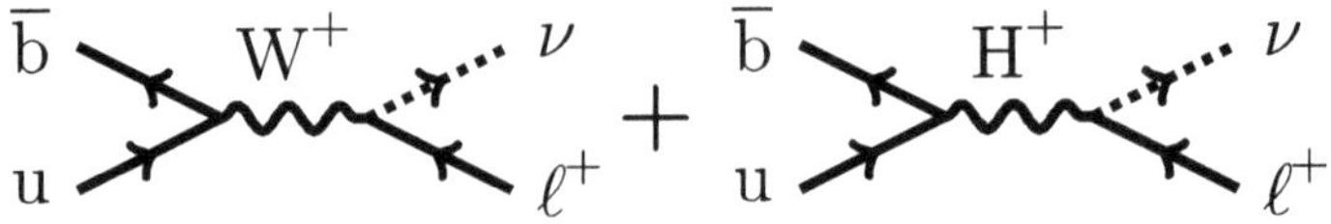

Fig. 5.2 Feynman Diagrams of the leading standard model process and a hypothetical new physics process mediated by a charged Higgs boson for the purely leptonic decay of a charged B meson

(amplitudes) interfere destructively, hence the branching fraction expected in the SM is further suppressed by a multiplicative factor [4, Sect. 7.10.2]:

$$\mathcal{B}\left(\mathrm{B}^+ \to \ell^+\nu\right)_{\mathrm{2HDM}} = \mathcal{B}\left(\mathrm{B}^+ \to \ell^+\nu\right)_{\mathrm{SM}} \cdot \left(1 - \frac{M_{\mathrm{B}}^2 \tan^2\beta}{M_{\mathrm{H}^+}^2}\right)^2 . \tag{5.2}$$

The mass of the charged Higgs M_{H^+} is already tightly constrained by weak radiative B meson decays $M_{\mathrm{H}^+} > 580\,\mathrm{GeV}$ [7]. Nevertheless, a valid value of $\tan\beta \geq 35$ can still yield a modification of the branching fraction, as large as the current theoretical and experimental uncertainties.

Therefore, the measurement of $\mathrm{B} \to \ell\nu$ can provide independent constraints on beyond the SM models like the Type II Two Higgs Doublet Model.

5.1.3 The $\mathrm{B} \to \tau\tau$ Puzzle

The decay $\mathrm{B} \to \tau\tau$ is extremely suppressed in the Standard Model [8]

$$\mathcal{B}(\mathrm{B}^0 \to \tau^-\tau^+)_{\mathrm{SM}} = (2.22 \pm 0.19) \cdot 10^{-8}. \tag{5.3}$$

Surprisingly, Belle measured the decay $\mathrm{B} \to \tau\tau$ and found a significant 5 sigma excess over the SM expectation [8]

$$\mathcal{B}(\mathrm{B}^0 \to \tau^-\tau^+)_{\mathrm{Belle}} = (4.39^{+0.80}_{-0.83} \pm 0.45) \cdot 10^{-3}. \tag{5.4}$$

The measurement used the `FR` for exclusive hadronic tagging and reconstructed the τ particles in one-prong decays. It was not accepted by the collaboration and not published in a peer-reviewed paper.

The BaBar experiment searched for the same decay and set an upper limit compatible with the Belle measurement (see [9])

$$\mathcal{B}(\mathrm{B}^0 \to \tau^-\tau^+)_{\mathrm{BaBar}} < 4.1 \cdot 10^{-3} \quad @90\%\mathrm{C.L.}. \tag{5.5}$$

The measurement used exclusive hadronic tagging and reconstructed the τ particles in one-prong decays. The exclusive hadronic tagging of BaBar relies on a different algorithm than used by Belle.

Finally, the LHCb experiment searched for the same decay and set an upper limit incompatible with the Belle measurement (see [10])

$$\mathcal{B}(\mathrm{B}^0 \to \tau^-\tau^+)_{\mathrm{LHCb}} < 2.1 \cdot 10^{-3} \quad @95\%\mathrm{C.L.}. \tag{5.6}$$

Since LHCb is not located at a B factory, the measurement did not use exclusive hadronic tagging. Instead a high statistics sample was used and the τ was reconstructed in three-prong decays.

The second-most[1] likely[2] explanation for these results is an unaccounted SM background process, which generates a peaking background contribution in one-prong, but not in three-prong decays.

The analyses strategies used by Belle and BaBar (exclusive tagging and one-prong decays) are very similar to the analyses presented in Sect. 5.1.5. In fact, the Belle measurement would indicate that $B^0 \to \tau^-\tau^+$ is a sizeable peaking background component in the measurement of $B^+ \to \tau^+\nu_\tau$.

The measurement of $B^+ \to \tau^+\nu_\tau$ using the new (independently developed) `BASF2` software applied to the $\Upsilon(4S)$ dataset recorded by Belle, can provide additional insight into this situation. In particular, the encountered large systematic uncertainty due to the continuum description could be a possible explanation (see Sect. 5.3).

5.1.4 τ Decay Models

This thesis investigates $B \to \tau\nu_\tau$, that is the leptonic decay of the charged B meson involving the heaviest charged lepton. This decay is particularly challenging because the τ is not stable in the detector, consequently the measurement has to deal with one or more neutrinos and the kinematics of the τ decay.

The kinematics of the decay $\tau \to \ell\nu\bar{\nu}$ is well-known.[3] The same holds true for $\tau \to \pi\nu$, where only a single hadron has to be considered.[4] The decay into multiple hadrons (in particular $\tau \to \pi\pi\pi\nu$) is more complicated due to the occurring intermediate resonances and their interference. Belle explicitly simulated the decays $\tau \to \rho[\to \pi\pi\nu]\nu$ and $\tau \to a_1[\to \pi\pi\pi]\nu$ without taking further decay channels into the same final state into account.[5] In contrast, Belle II simulates the hadronic decay of $\tau \to \pi\pi\nu$ and $\tau \to \pi\pi\pi\nu$ inclusively based on [11] and takes interferences into account.[6]

The implications of this different treatment in Belle and Belle II for this analysis are mostly technical details related to the Monte Carlo matching algorithm, and are not further discussed.

5.1.5 Previous Measurements

The decay $B \to \tau\nu_\tau$ was already observed in previous measurements at B factories. Both, the Belle [1, 2] and the BaBar [12, 13] experiment conducted measurements

---

[1]The most likely explanation is a technical error in the Belle measurement.

[2]In a Bayesian sense.

[3]The `EvtTaulnunu` model of `evtgen` is used.

[4]The `EvtTauScalarnu` model of `evtgen` is used.

[5]The `EvtTauVectornu` model of `evtgen` is used.

[6]The `EvtTauHadnu` model of `evtgen` is used.

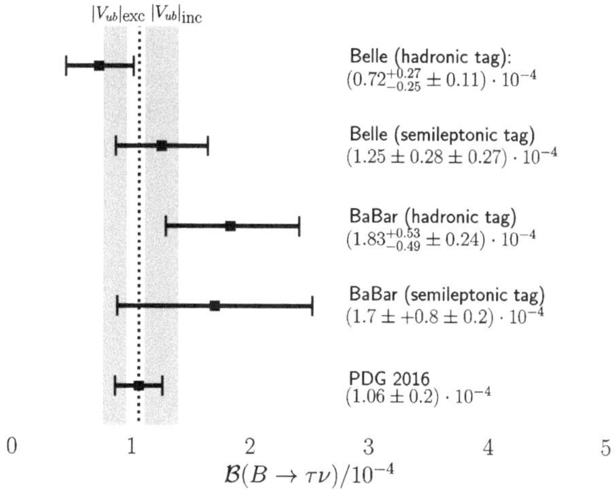

Fig. 5.3 Current experimental status of the measurement, including the standard model predictions (gray bands) using V_{ub} from exclusive and inclusive measurements

using hadronic and semileptonic tagging. The previous measurements and the Standard Model predictions using V_{ub} from exclusive (SMExc) and inclusive (SMInc) measurements are shown in Fig. 5.3. The Belle measurements are more precise due to the larger recorded dataset. The hadronic and semileptonic tags have similar statistical power. The current world average of the measured branching fraction stated by [5], is compatible with both predictions SMExc and SMInc.

5.2 Measurement

The measurement of the branching fraction of the decay B $\to \tau\nu_\tau$ involves five reconstruction steps, which are explained in detail in this chapter.

1. The events are preselected (see Sect. 5.2.1).
2. The signal side is reconstructed in five distinct decay channels of the τ lepton (see Sect. 5.2.2).
3. The tag side is reconstructed by the `FEI`, which provides a hadronic and a semileptonic tag (see Sect. 5.2.3).
4. Both sides are combined to a $\Upsilon(4S)$ candidates and the **completeness-constraint** is applied (see Sect. 5.2.4).
5. The branching fraction is extracted with an extended unbinned maximum likelihood fit on the extra energy in the electromagnetic calorimeter of the selected candidates (see Sect. 5.2.5).

The measurement is performed on converted data recorded by the Belle detector. The hadronically and semileptonically tagged samples are treated as independent measurements in this thesis. The hadronic tag already includes the tag-side efficiency correction factors obtained by [14]. At the time of writing there was no tag-side efficiency correction available for the semileptonic tag. In consequence, the semileptonic results have to be taken with a grain of salt.

If not stated otherwise, all figures and tables were produced using one stream ($\equiv$711 fb^{-1}) of Monte Carlo simulated Belle events to model the background processes and a high-statistics sample of 100 million Monte Carlo simulated signal-events.

5.2.1 Skimming

The measurements consider only events with ten or less **good tracks**, that is tracks with impact parameters of less than 2 cm and 4 cm in transverse and beam axis direction, respectively.

This reduces the computational effort drastically without loosing any signal events. As was shown in Fig. 4.13, the fraction of B_{tag} candidates with more than seven tracks, which are correctly tagged by the `FEI`, is negligible. The B_{sig} requires at most three tracks (for most channels only one). Therefore, the `FEI` cannot assign a correct tag to signal events with more than ten good tracks, and the events can be discarded without loosing correctly tagged signal events. Even incorrectly tagged signal events are only affected in the three-prong τ decay channel, as can be seen from Fig. 4.13b.

Figure 5.4 shows the distribution of the number of tracks and the corresponding selection efficiencies. The observed difference between Monte Carlo simulation and data was already investigated and discussed in Sect. 2.3.3.3.

5.2.2 Signal Side Reconstruction

The B meson travels less than 1 mm inside the detector due to its short life-time of $1.64 \cdot 10^{-12}$ s. The τ lepton, produced by the investigated decay $B \rightarrow \tau\nu_\tau$, has an even shorter life-time of $2.9 \cdot 10^{-13}$ s. Therefore the τ lepton does not leave the beam-pipe and cannot be detected directly.

In this thesis the signal-side is reconstructed in five different τ decay-channels covering a total branching fraction of $\mathcal{BR}(\tau) = 80.8\%$ (see Fig. 5.5).

Each decay-channel has distinct properties and consequently individual selection criteria, which are described below. As can be seen from Table 5.3, the signal-side selection criteria suppress the background by two orders of magnitude, while maintaining most of the signal. The measured selection efficiencies on converted data are compatible with the Monte Carlo simulation of background.

Although the tag-side is reconstructed by the `FEI`, it proved useful to already discard candidates whose rest-of-event is not compatible with a single B meson. A fast, efficient and easy to calculate selection variable is the deviation ΔE of the total energy in the center-of-mass system of the rest-of-event from the nominal beam-

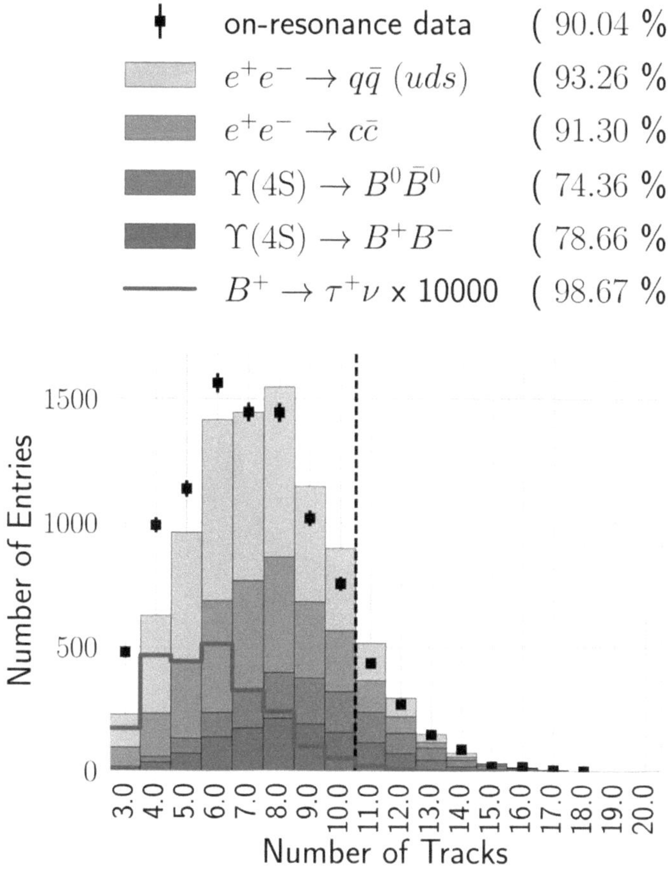

Fig. 5.4 Skim selection on the number of good tracks. Evaluated on 10,000 events for each component, scaled according to their expected relative fractions

Table 5.3 The overall efficiencies of the signal-side selection: $\varepsilon_{\text{SignalEvents}}$, is the fraction of events containing the stated signal decay channel, in which a correct candidate was reconstructed. This includes the detector acceptance, trigger and reconstruction efficiencies and the skimming. The signal-side selection alone is characterized by: the fraction of correct candidates which survive the signal-side selection $\varepsilon_{\text{Signal}}$; the fraction of background candidates which survive the signal-side selection $\varepsilon_{\text{Background}}$; and the fraction of candidates on converted data which survive the signal-side selection $\varepsilon_{\text{Data}}$. Evaluated on 10,000 events for each component

Channel	$\varepsilon_{\text{SignalEvents}}$	$\varepsilon_{\text{Signal}}$	$\varepsilon_{\text{Background}}$	$\varepsilon_{\text{Data}}$
μ	0.40	0.48	0.006	0.006
e	0.68	0.71	0.008	0.010
π	0.89	0.88	0.087	0.096
ρ	0.41	0.65	0.014	0.015
a_1	0.44	0.67	0.060	0.058

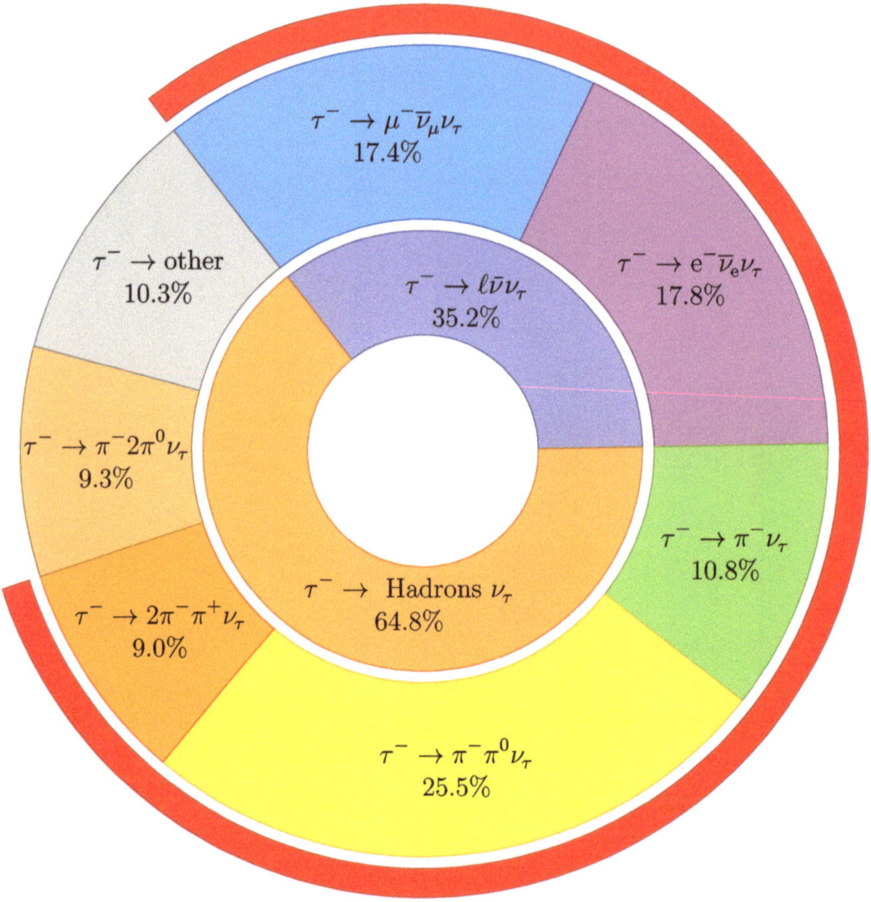

Fig. 5.5 Visualization of the τ branching fractions: used in this thesis (red arc); into exclusive decay-channels (outer circle); and into inclusive hadronic and leptonic decay-channels (inner circle)

energy in the center-of-mass system. Figure 5.6 shows the efficiency of this selection for the $\tau \to \mu\nu\bar{\nu}$ decay channel, the distributions in the other channels look nearly identical. This selection saves computation time. Once the tag-side is reconstructed by the `FEI` a superior[7] selection on the same quantity can be performed.

5.2.2.1 $\tau \to \mu\nu\bar{\nu}$

The τ decay into a muon and two neutrinos has a branching fraction of 17.4%. It can be efficiently selected using the muon identification provided by the PID sub-

[7]The knowledge of the explicit decay-chain of the B_{tag} increases the resolution of ΔE.

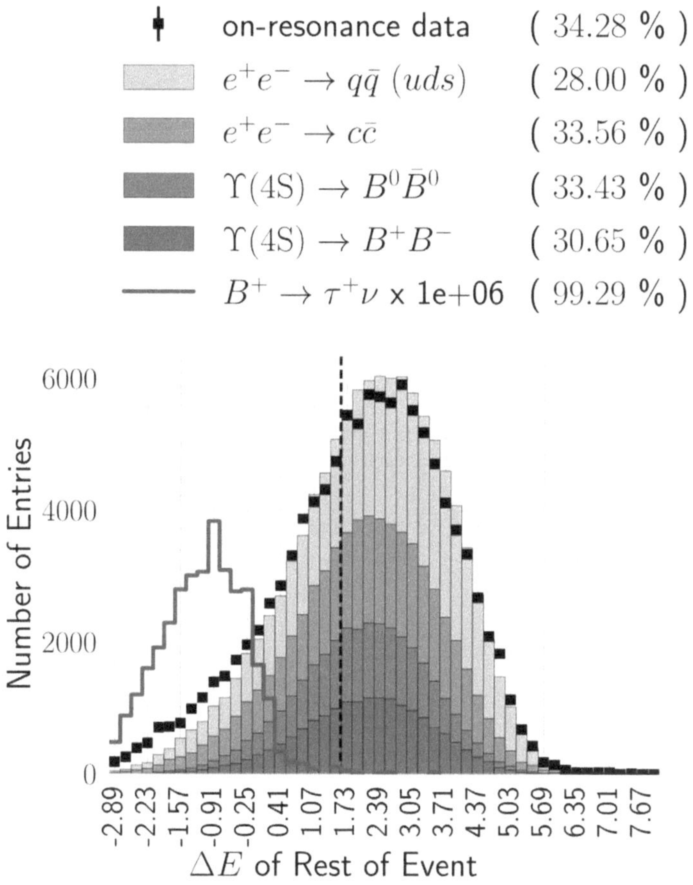

Fig. 5.6 ΔE distribution of the rest-of-event in GeV for the $\tau \rightarrow \mu\nu\bar{\nu}$ decay channel with a cut on $\Delta E < 1.5$ GeV. Evaluated on 10,000 events for each component, scaled according to their expected relative fractions

detectors. We expect to find one track compatible with a muon hypothesis in the event besides the B_{tag} meson. The selection criteria and corresponding efficiencies are stated in Table 5.4.

5.2.2.2 $\tau \rightarrow e\nu\bar{\nu}$

The τ decay into an electron and two neutrinos has a branching fraction of 17.8%. It can be efficiently selected using the electron identification provided by the PID sub-detectors. We expect to find one track compatible with an electron hypothesis in the event besides the B_{tag} meson. The selection criteria and corresponding efficiencies are stated in Table 5.5.

Table 5.4 The selection criteria for the $\tau \rightarrow \mu\nu\bar{\nu}$ decay-channel with the corresponding selection efficiencies for signal candidates $\varepsilon_{\text{Signal}}$, background candidates $\varepsilon_{\text{Background}}$ and for candidates on data $\varepsilon_{\text{Data}}$. Evaluated on 10,000 events for each component

Selection criterion	$\varepsilon_{\text{Signal}}$	$\varepsilon_{\text{Background}}$	$\varepsilon_{\text{Data}}$
# Tracks ≤ 10	0.99	0.82	0.84
$p_\tau^* > 0.1\,\text{GeV}$	1.00	0.97	0.97
$\Delta E \leq 1.5\,\text{GeV}$	0.99	0.31	0.34
dr < 2	0.98	0.75	0.74
$-4 <$ dz < 4	0.96	0.68	0.66
PID- > 0.9	0.49	0.01	0.01

Table 5.5 The selection criteria for the $\tau \rightarrow e\nu\bar{\nu}$ decay-channel with the corresponding selection efficiencies for signal candidates $\varepsilon_{\text{Signal}}$, background candidates $\varepsilon_{\text{Background}}$ and for candidates on data $\varepsilon_{\text{Data}}$. Evaluated on 10,000 events for each component

Selection criterion	$\varepsilon_{\text{Signal}}$	$\varepsilon_{\text{Background}}$	$\varepsilon_{\text{Data}}$
# Tracks ≤ 10	0.99	0.82	0.84
$p_\tau^* > 0.1\,\text{GeV}$	0.99	0.97	0.97
$\Delta E \leq 1.5\,\text{GeV}$	0.99	0.31	0.34
dr < 2	0.96	0.75	0.74
$-4 <$ dz < 4	0.94	0.68	0.66
PID- < 0.9	1.00	0.99	0.99
$\text{PID}_\text{e} > 0.9$	0.74	0.04	0.04

5.2.2.3 $\tau \rightarrow \pi\nu_\tau$

The τ decay into a pion and one neutrino has a branching fraction of 10.8%. It is a two-body decay, hence the momentum of the pion in the rest-frame of the τ is known

$$p_\pi^* = \frac{1}{2}\left(m_\tau - \frac{m_\pi^2}{m_\tau}\right) = 0.88\,\text{GeV}. \tag{5.7}$$

Together with the momentum of the τ lepton in the rest-frame of the B meson and the momentum of the B meson in the center-of-mass system

$$p_\tau^* = 2.34\,\text{GeV} \tag{5.8}$$

$$p_\text{B}^* = 0.33\,\text{GeV} \tag{5.9}$$

we can calculate the momentum of the pion in the center-of-mass system using the rapidities of the τ lepton and B^+ meson

$$\eta_\tau^* = \text{arsinh}\frac{p_\tau^*}{m_\tau} = 1.089 \tag{5.10}$$

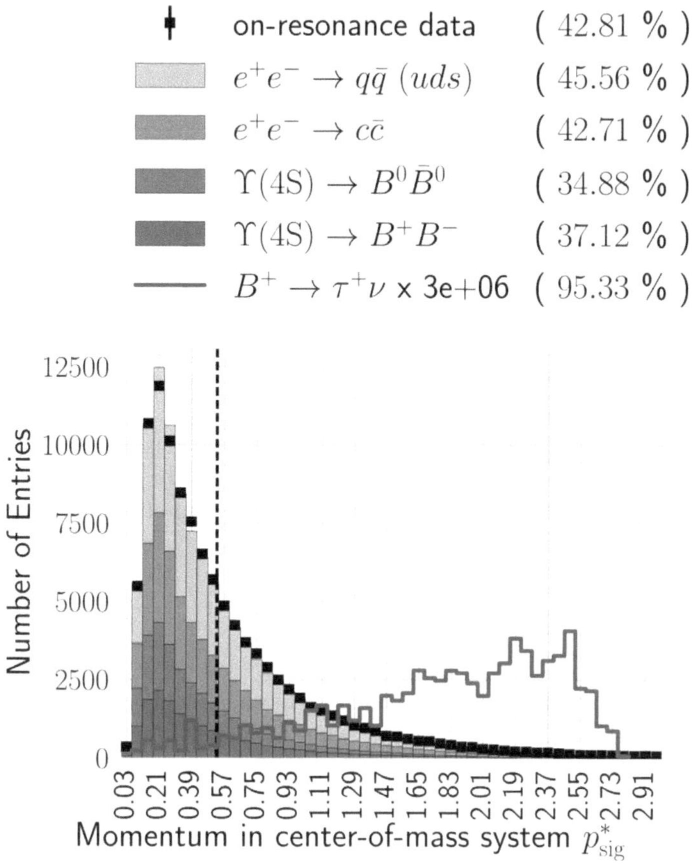

Fig. 5.7 p^*_{sig} distribution in GeV for the $\tau \to \pi\nu_\tau$ decay channel with a cut on $p^*_{\mathrm{sig}} < 0.5\,\mathrm{GeV}$. Evaluated on 10,000 events for each component, scaled according to their expected relative fractions

$$\eta^*_{\mathrm{B}} = \mathrm{arsinh}\frac{p^*_{\mathrm{B}}}{m_{\mathrm{B}}} = 0.062 \tag{5.11}$$

and finally the minimal and maximal pion momentum in the center-of-mass system

$$\eta = \pm\eta^*_\tau \pm \eta^*_{\mathrm{B}} \tag{5.12}$$

$$p_\pi = p^*_\pi \cosh\eta - \sqrt{m^2_\pi + (p^*_\pi)^2}\sinh\eta \tag{5.13}$$

$$= 0.26\,\mathrm{GeV} \quad \text{minimum} \tag{5.14}$$

$$= 2.80\,\mathrm{GeV} \quad \text{maximum.} \tag{5.15}$$

The exact p_π distribution is skewed towards higher momenta due to the fixed helicity of the neutrino (see Fig. 5.7). We expect to find one track with a compatible

Table 5.6 The selection criteria for the $\tau \rightarrow \pi\nu_\tau$ decay-channel with the corresponding selection efficiencies for signal candidates $\varepsilon_{\text{Signal}}$, background candidates $\varepsilon_{\text{Background}}$ and for candidates on data $\varepsilon_{\text{Data}}$. Evaluated on 10,000 events for each component

Selection criterion	$\varepsilon_{\text{Signal}}$	$\varepsilon_{\text{Background}}$	$\varepsilon_{\text{Data}}$
# Tracks ≤ 10	0.99	0.76	0.82
$p_\tau^* > 0.5\,\text{GeV}$	0.95	0.42	0.43
$\Delta E \leq 1.5\,\text{GeV}$	0.99	0.31	0.34
dr < 2	0.96	0.75	0.74
$-4 <$ dz < 4	0.93	0.68	0.66
PID- < 0.9	0.99	0.99	0.99
$\text{PID}_\text{e} < 0.9$	1.00	0.96	0.96
$\text{PID}_\text{K} < 0.9$	0.97	0.88	0.87

momentum in the event besides the B_{tag} meson. In addition, we expect a large cross-feed in this decay-channel from other signal decay-channels. For instance, a lepton which does not pass the PID criteria will automatically be picked up by the pion decay-channel. The selection criteria and corresponding efficiencies are stated in Table 5.6.

5.2.2.4 $\tau \rightarrow \rho\nu_\tau$

The τ decay into a ρ meson and one neutrino has a branching fraction of 25.5%. It is the dominant decay channel of the τ with the largest branching fraction. The ρ resonance decays instantaneous into a charged and a neutral pion $\rho^+ \rightarrow \pi^+\pi^0[\rightarrow \gamma\gamma]$. The intermediate resonances ρ and π^0 provide efficient selection criteria based on their invariant masses

$$m_\rho = 0.775\,\text{GeV} \tag{5.16}$$

$$m_{\pi^0} = 0.135\,\text{GeV}. \tag{5.17}$$

We expect to find one track and two ECL clusters compatible with a π^0 (see Fig. 5.8) and a ρ (see Fig. 5.9) in the event besides the B_{tag} meson. The selection criteria and corresponding efficiencies are stated in Table 5.7.

5.2.2.5 $\tau \rightarrow \text{a}_1\nu_\tau$

The τ decay into three pions $\tau \rightarrow 3\pi$ is dominated by the decay-chain $\tau^+ \rightarrow \text{a}_1^+[\rightarrow \rho^0\pi^+]\nu_\tau$ [15]. In this thesis, the τ decay into three charged pions (via $\rho^0 \rightarrow \pi^+\pi^-$) was investigated for the first time, it has a branching fraction of 9.5%. The other possible decay into a charged pion and two neutral pions (via $\rho^0 \rightarrow \pi^0\pi^0$) was not studied.

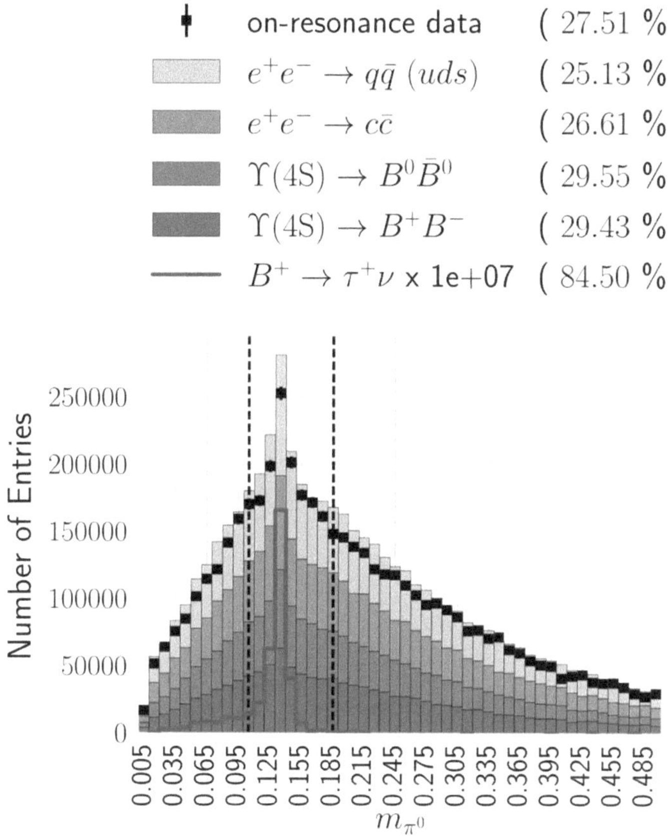

Fig. 5.8 M_{π^0} distribution in GeV for the $\tau \to \rho\nu_\tau$ decay channel with a cut on $0.1\,\text{GeV} < M_{\pi^0} < 0.18\,\text{GeV}$. Evaluated on 10,000 events for each component, scaled according to their expected relative fractions

The resonance a_1 provides an efficient selection criteria based on its invariant mass (see Fig. 5.10)

$$m_{a_1} = 1.230\,\text{GeV}. \tag{5.18}$$

We expect to find three tracks from a common vertex compatible with the expected intermediate resonances in the event besides the B_{tag} meson. This is the only signal decay-channel with a different multiplicity, that is number of tracks in the final state. Therefore it has distinct physics background contributions compared to all other studied decay-channels. Unfortunately, it also suffers from a large combinatorial background, which is the reason why it was not used in previous analyses. The selection criteria and corresponding efficiencies are stated in Table 5.8.

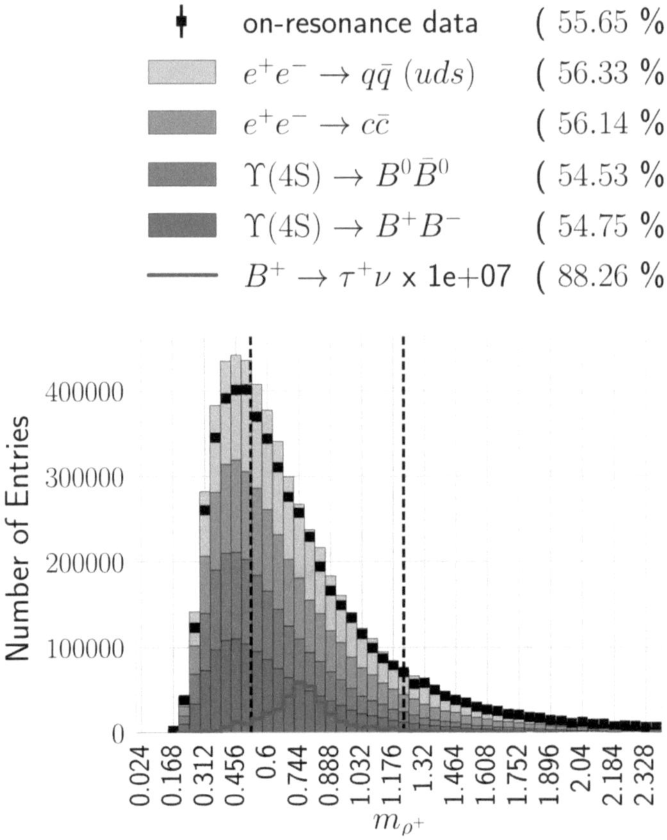

Fig. 5.9 M_ρ distribution in GeV for the $\tau \to \rho\nu_\tau$ decay channel with a cut on 0.5 GeV $< M_\rho <$ 1.2 GeV. Evaluated on 10,000 events for each component, scaled according to their expected relative fractions

5.2.3 Tag Side Reconstruction

The default generic FEI (see Chap. 4) applied to the entire event was used to reconstruct the tag-side of the event. The FEI provides B meson tag-side candidates in **hadronic** and **semileptonic** decay channels. Both were used in this analysis. Due to the different systematic uncertainties (in particular the missing calibration factors for the semileptonic tag), the tags are treated as independent measurements in this thesis, that is in principle the same event could appear in both tagged samples.

The default FEI does not use information from the M_{bc} and $\cos\Theta_{\text{BD}\ell}$ of the hadronically and semileptonically tagged B meson, respectively. In addition the FEI aims to always output a valid candidate, even if the associated Signal Probability σ is very low. However, this analysis requires a high-purity tag, because the signal-side selection alone does not provide enough separation power.

Table 5.7 The selection criteria for the $\tau \rightarrow \rho\nu_\tau$ decay-channel with the corresponding selection efficiencies for signal candidates $\varepsilon_{\text{Signal}}$, background candidates $\varepsilon_{\text{Background}}$ and for candidates on data $\varepsilon_{\text{Data}}$. Evaluated on 10,000 events for each component

Selection criterion	$\varepsilon_{\text{Signal}}$	$\varepsilon_{\text{Background}}$	$\varepsilon_{\text{Data}}$
# Tracks ≤ 10	0.99	0.81	0.82
$p_\tau^* > 0.6\,\text{GeV}$	0.96	0.45	0.46
$\Delta E \leq 1.5\,\text{GeV}$	1.00	0.35	0.37
dr < 2	0.93	0.74	0.72
−4 < dz < 4	0.88	0.65	0.64
PID- < 0.9	1.00	0.99	0.99
$\text{PID}_\text{e} < 0.9$	0.99	0.96	0.95
$\text{PID}_\text{K} < 0.9$	0.98	0.88	0.87
$0.5\,\text{GeV} < M_\rho < 1.2\,\text{GeV}$	0.88	0.56	0.56
$0.1\,\text{GeV} < M_{\pi^0} < 0.18\,\text{GeV}$	0.84	0.27	0.28

Therefore the tag-side candidates have to fulfill the selection criteria stated in Table 5.9 to be considered in the analysis.

5.2.4 Event Reconstruction

The selected signal-side and tag-side candidates are combined to an $\Upsilon(4S)$ candidate.

5.2.4.1 Completeness-Constraint(s)

The completeness-constraint is applied and all candidates with additional good tracks in the event (besides the ones, which were used for the reconstruction of the $\Upsilon(4S)$) are discarded. The completeness-constraint is the most-effective selection criterion in this analysis. It has a very high signal efficiency, while it reduces the background by one order of magnitude in all signal-side (the five τ decay-channels) and tag-side (hadronic and semileptonic) combinations. The efficiencies are stated in Table 5.10. An example distribution for the $\tau \rightarrow \pi\nu_\tau$ is shown in Fig. 5.11.

The completeness-constraint is part of an entire class of constraints on additional physics objects in the event besides the reconstructed $\Upsilon(4S)$. Conveniently, the `FEI` provides candidates with an associated `SignalProbability` for many particle

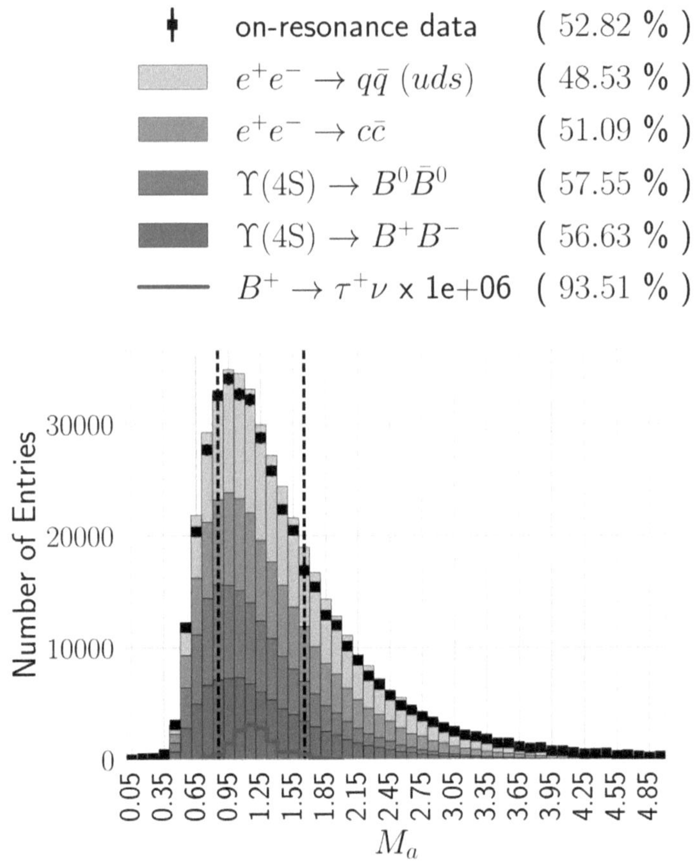

Fig. 5.10 M_{a_1} distribution in GeV for the $\tau \to a_1\nu_\tau$ decay channel with a cut on $0.8\,\text{GeV} < M_{a_1} < 1.6\,\text{GeV}$. Evaluated on 10,000 events for each component, scaled according to their expected relative fractions

Table 5.8 The selection criteria for the $\tau \to a_1\nu_\tau$ decay-channel with the corresponding selection efficiencies for signal candidates $\varepsilon_{\text{Signal}}$, background candidates $\varepsilon_{\text{Background}}$ and for candidates on data $\varepsilon_{\text{Data}}$. The impact parameter selection criteria are the same for each of the three tracks. Evaluated on 10,000 events for each component

Selection criterion	$\varepsilon_{\text{Signal}}$	$\varepsilon_{\text{Background}}$	$\varepsilon_{\text{Data}}$
# Tracks ≤ 10	0.94	0.64	0.66
$p^*_\tau > 0.9\,\text{GeV}$	0.96	0.51	0.52
$\Delta E \leq 1.5\,\text{GeV}$	1.00	0.53	0.55
dr < 2	0.99	0.96	0.95
$-4 <$ dz < 4	0.97	0.90	0.88
$\chi^2_{\text{Vertex}} > 0.01$	0.79	0.53	0.50
$0.8\,\text{GeV} < M_{a_1} < 1.6\,\text{GeV}$	0.94	0.53	0.53

Table 5.9 The selection criteria for the tagged B mesons with the corresponding selection efficiencies for correctly reconstructed candidates $\varepsilon_{\text{Signal}}$, background candidates $\varepsilon_{\text{Background}}$ and for candidates on data $\varepsilon_{\text{Data}}$. Evaluated on 10 million events for each component

Selection criterion	$\varepsilon_{\text{Signal}}$	$\varepsilon_{\text{Background}}$	$\varepsilon_{\text{Data}}$
Hadronic tag			
$M_{bc} \geq 5.27$	0.98	0.121	0.117
$-0.15 < \Delta E < 0.1$	0.97	0.593	0.572
$\sigma > 0.005$	0.94	0.155	0.147
Total	0.91	0.025	0.024
Semileptonic tag			
$-1 < \cos\Theta_{BD\ell} < 1$	0.88	0.293	0.280
$\sigma > 0.005$	0.90	0.130	0.126
Total	0.80	0.053	0.050

Table 5.10 Completeness-constraint on the number of additional good tracks. Stated are the selection efficiencies for: correctly reconstructed candidates $\varepsilon_{\text{Signal}}$, background candidates $\varepsilon_{\text{Background}}$, and for candidates on data $\varepsilon_{\text{Data}}$. Evaluated on 10 million events for each component

Decay-channel	$\varepsilon_{\text{Signal}}$	$\varepsilon_{\text{Background}}$	$\varepsilon_{\text{Data}}$
Hadronic tag			
e^-	0.96	0.11	0.11
μ^-	0.96	0.12	0.13
π	0.96	0.11	0.11
ρ	0.95	0.12	0.14
a_1	0.96	0.17	0.17
Semileptonic tag			
e^-	0.97	0.05	0.05
μ^-	0.97	0.05	0.06
π	0.97	0.03	0.03
ρ	0.95	0.03	0.03
a_1	0.95	0.05	0.05

types. In particular, completeness-constraint like selections on the number of good[8] additional photons, K_S^0 and π^0 can be performed. The cut on additional photons is strongly correlated to the final fit variable E_{ECL}, hence only the cuts on the additional K_S^0 and π^0 are applied. The efficiencies are stated in Table 5.11 for π^0 and in Table 5.12 for K_S^0. An example distribution for the $\tau \to \pi\nu_\tau$ is shown in Fig. 5.12 for π^0 and in Fig. 5.13 for K_S^0.

[8]Candidates with a `SignalProbability` larger than 0.1 are considered good.

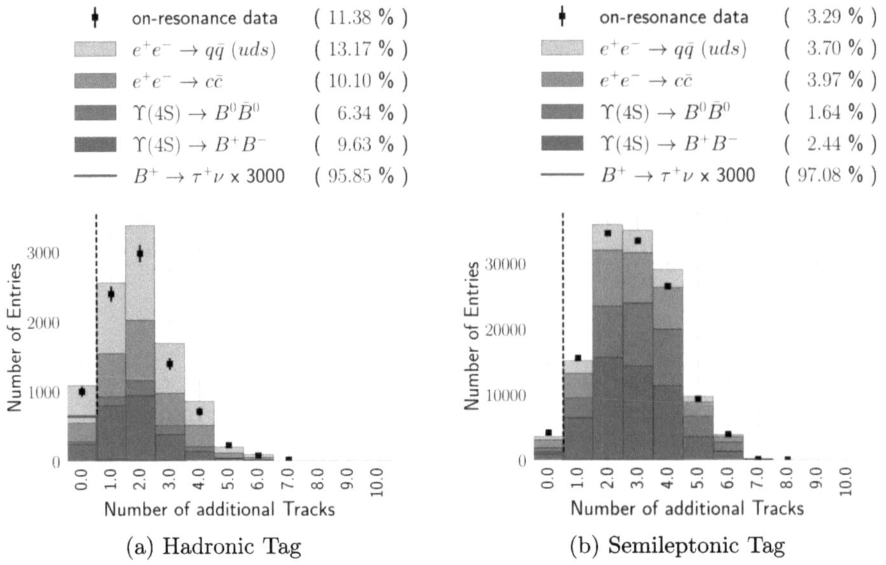

Fig. 5.11 Distribution of the number of additional good tracks in the event not used for the $\Upsilon(4S)$ reconstruction for the $\tau \to \pi\nu_\tau$ decay channel. Evaluated on 10 million events for each component, scaled according to their expected relative fractions

Table 5.11 Completeness-constraint on the number of additional π^0. Stated are the selection efficiencies for: correctly reconstructed candidates $\varepsilon_{\text{Signal}}$, background candidates $\varepsilon_{\text{Background}}$, and for candidates on data $\varepsilon_{\text{Data}}$. Evaluated on 10 million events for each component

Decay-channel	$\varepsilon_{\text{Signal}}$	$\varepsilon_{\text{Background}}$	$\varepsilon_{\text{Data}}$
Hadronic tag			
e^-	0.83	0.42	0.38
μ^-	0.85	0.39	0.42
π	0.82	0.36	0.34
ρ	0.86	0.33	0.32
a_1	0.79	0.33	0.34
Semileptonic tag			
e^-	0.84	0.27	0.27
μ^-	0.87	0.28	0.27
π	0.84	0.21	0.21
ρ	0.88	0.19	0.18
a_1	0.79	0.18	0.18

Table 5.12 Completeness-constraint on the number of additional K_S^0. Stated are the selection efficiencies for: correctly reconstructed candidates $\varepsilon_{\text{Signal}}$, background candidates $\varepsilon_{\text{Background}}$, and for candidates on data $\varepsilon_{\text{Data}}$. Evaluated on 10 million events for each component

Decay-channel	$\varepsilon_{\text{Signal}}$	$\varepsilon_{\text{Background}}$	$\varepsilon_{\text{Data}}$
Hadronic tag			
e^-	1.00	0.93	0.91
μ^-	1.00	0.93	0.91
π	1.00	0.92	0.92
ρ	1.00	0.93	0.93
a_1	1.00	0.94	0.33
Semileptonic tag			
e^-	1.00	0.90	0.90
μ^-	1.00	0.90	0.92
π	1.00	0.89	0.90
ρ	1.00	0.90	0.90
a_1	1.00	0.92	0.92

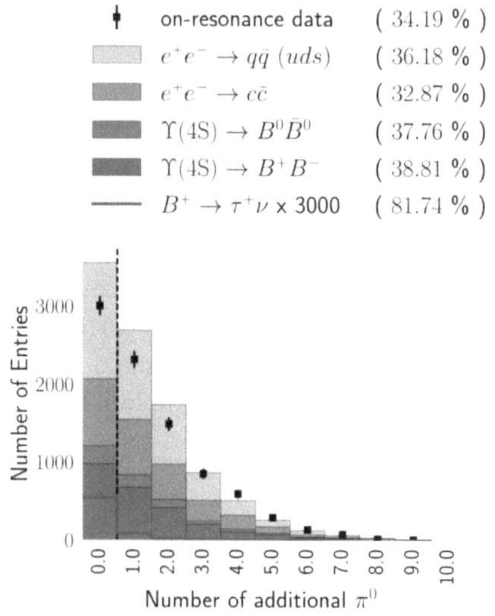

(a) Hadronic Tag

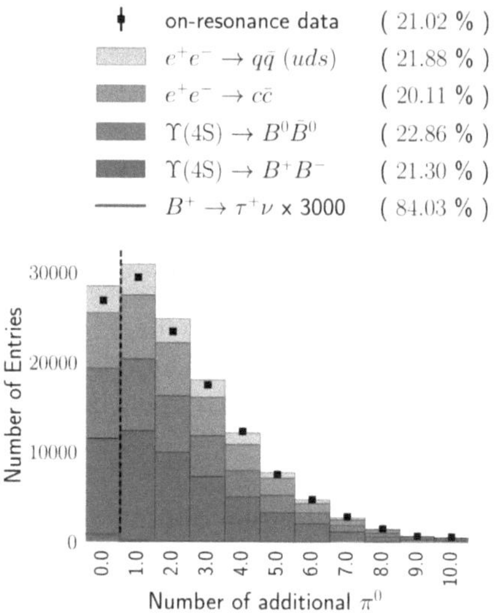

(b) Semileptonic Tag

Fig. 5.12 Distribution of the number of additional good π^0 in the event not used for the $\Upsilon(4S)$ reconstruction for the $\tau \to \pi\nu_\tau$ decay channel. Evaluated on 10 million events for each component, scaled according to their expected relative fractions

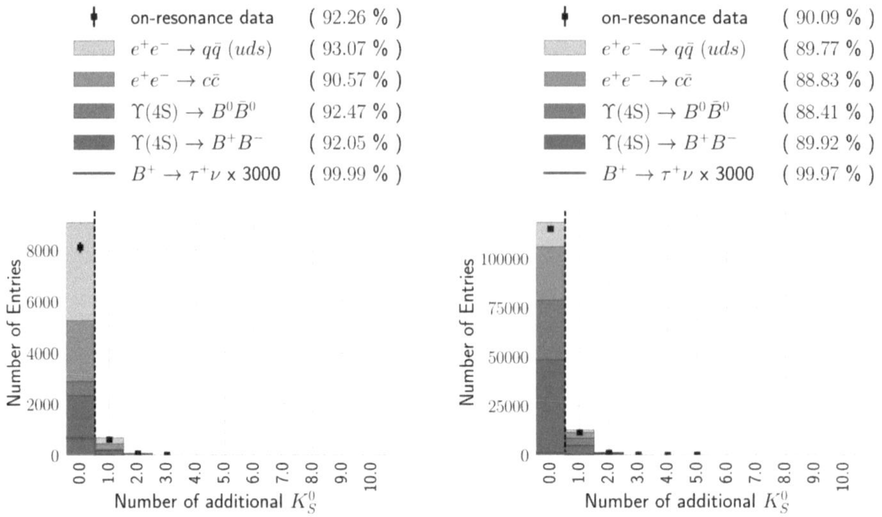

(a) Hadronic Tag (b) Semileptonic Tag

Fig. 5.13 Distribution of the number of additional good K_S^0 in the event not used for the $\Upsilon(4S)$ reconstruction for the $\tau \to \pi\nu_\tau$ decay channel. Evaluated on 10 million events for each component, scaled according to their expected relative fractions

5.2.4.2 Best-Candidate Selection

The chosen signal-side selection criteria and the completeness-constraint ensure that all signal-side candidates are (nearly[9]) mutual exclusive. Therefore a best-candidate selection solely based on the `SignalProbability` of the tag-side is performed. Hence, the same event can only contribute to one of the five investigated decay-modes. The efficiencies of the best-candidate selection are stated in Table 5.13.

5.2.4.3 Continuum Suppression

In previous and similar Belle analyses a continuum suppression algorithm based on a multivariate method using event shape features was used to reduce the background from continuum processes (see Sect. 3.3.1.3).

In this benchmark analysis, I take a more conservative approach to prevent any systematic effects caused by an additional multivariate method besides the `FEI`.

The continuum events are suppressed by a cut on a single well-understood variable

$$T = |\cos(\sphericalangle(T_{\text{tag}}, T_{\text{sig}}))|, \tag{5.19}$$

[9]In rare cases a valid π and ρ candidate can be reconstructed in the same event, where the π^0 from the ρ is not considered a good π^0 from the `FEI`. In these cases the ρ candidate is discarded in favor of the π candidate.

Table 5.13 Best-candidate selection efficiencies for correctly reconstructed candidates $\varepsilon_{\text{Signal}}$, background candidates $\varepsilon_{\text{Background}}$ and for candidates on data $\varepsilon_{\text{Data}}$. The efficiencies are calculated after all completeness-constraints were applied. The statistical power of the background and data sample was very limited, hence the stated efficiencies have large uncertainties ± 0.1. Evaluated on 10 million events for each component

Decay-channel	$\varepsilon_{\text{Signal}}$	$\varepsilon_{\text{Background}}$	$\varepsilon_{\text{Data}}$
Hadronic tag			
e^-	0.86	1.0	1.0
μ^-	0.86	1.0	0.9
π	0.86	0.9	0.9
ρ	0.81	0.8	0.7
a_1	0.87	1.0	1.0
Semileptonic tag			
e^-	0.97	1.0	1.0
μ^-	0.97	1.0	1.0
π	0.96	0.9	0.9
ρ	0.91	0.8	0.8
a_1	0.97	1.0	1.0

where T_{tag} and T_{sig} denote the thrust axis [4, Sect. 9.3] of the tag-side and signal-side, respectively. Correctly reconstructed $\Upsilon(4S)$ events have a flat distribution in this variable, because the two B meson decay spherically and independent of one another. Hence, the signal selection efficiency is uniform, and no additional systematic uncertainty is introduced. On the other hand, continuum events peak at large values of T, because of their jet-like decay topology. This theoretical assumption is verified (also for incorrectly reconstructed $\Upsilon(4S)$ decays) using Monte Carlo simulation by Fig. 5.14.

5.2.4.4 Final Selection

A final selection optimized for each tag-side and signal-side combination separately is performed using: the `SignalProbability` σ of the tag-side, and the continuum suppression variable T.

In contrast to the previous selections, this final selection is performed during the **n-tuple analysis**. A multi-dimensional minimization algorithm optimizing the cuts by minimizing the expected negative $\frac{S}{S+B}$ was investigated. However, this approach was not used in the final analysis, because the optimum is located in a problematic phase-space region, with peaking background contributions from continuum processes not present in the simulated data (see Sect. 5.3). Instead, the cuts were chosen in order to avoid those problematic phase-space regions.

A loose cut on the `SignalProbability` $\sigma > 0.01$ was chosen for the hadronically tagged channels, whereas a tighter cut $\sigma > 0.05$ was chosen for the

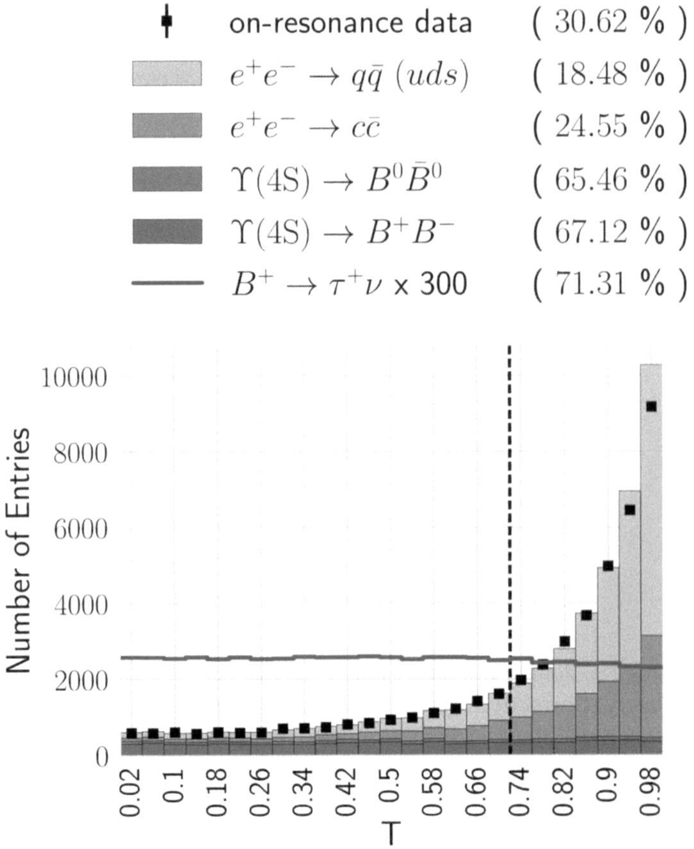

Fig. 5.14 Distribution of the angle between the thrust axes of the signal-side and tag-side for the hadronically tagged $\tau \to \pi\nu_\tau$ decay channel. The deviations between Monte Carlo simulation and data is discarded by a cut at 0.7

semileptonically tagged channels. This reflects the difference in the purity of the samples provided by the respective tags. The hadronic τ decay modes suffer from large continuum backgrounds, a tight cut $T < 0.7$ was chosen, to suppress those continuum events. In particular this selection discards the phase space region with the largest data/Monte Carlo disagreement. In contrast, the leptonic τ decay modes are expected to have only a small continuum contribution, nevertheless a loose cut $T < 0.85$ was applied to ensure that peaking contributions from unknown continuum processes are discarded (see Sect. 5.3.2.1).

The final selection efficiencies and the expected number of events, as well as the expected significance of a counting experiment of all events with an E_{ECL} below 100 MeV are stated in Table Table 5.14 and Table 5.15

Previous analyses discarded in addition all events with additional non-good tracks. Although this selection is very effective and suppresses additional background it was not applied in this analyses for the following reasons. The non-good tracks are not

Table 5.14 Final cuts on T and σ. Stated are the selection efficiencies for; correctly reconstructed candidates $\varepsilon_{\text{Signal}}$, background candidates $\varepsilon_{\text{Background}}$, and for candidates on data $\varepsilon_{\text{Data}}$. $\frac{S}{\sqrt{B}}$ refers to the expected significance of a counting experiment of all events (number of signal events S and number of background events B) below 100 MeV

Decay-channel	$\frac{S}{\sqrt{B}}$	$\varepsilon_{\text{Signal}}$	$\varepsilon_{\text{Background}}$	$\varepsilon_{\text{Data}}$
Hadronic tag				
e^-	2.40	0.73	0.39	0.39
μ^-	1.77	0.73	0.40	0.40
π	2.39	0.59	0.13	0.13
ρ	1.16	0.56	0.11	0.11
a_1	0.37	0.57	0.11	0.11
Semileptonic tag				
e^-	2.12	0.35	0.10	0.09
μ^-	1.49	0.36	0.10	0.10
π	2.23	0.28	0.04	0.04
ρ	1.43	0.24	0.03	0.03
a_1	0.40	0.24	0.04	0.04

Table 5.15 Expected number of events from the Monte Carlo simulation and the obtained number on on-resonance and off-resonance data. The uncertainty of the stated numbers is below 2%

	Hadronic					Semileptonic				
	e	μ	π	ρ	a_1	e	μ	π	ρ	a_1
Signal	80	52	124	79	21	79	48	115	89	23
Background	4428	4056	6154	12,244	6915	3961	3382	4223	8032	5135
Total	4508	4109	6278	12,323	6937	4041	3430	4339	8122	5158
On-resonance	4079	3545	6019	11,572	5542	4131	3421	5070	9109	5050
Continuum	264	180	2625	5481	2685	61	43	1089	1604	720
Off-resonance	222	175	2306	5105	1702	135	103	1574	2059	707

well understood, they depend strongly on the beam-conditions and the reconstruction software (see Sect. 2.3.3.2). Furthermore the remaining off-resonance data sample after this selection does not allow a sound estimation of possible peaking background components. Previous analyses did not consider this potentially large systematic uncertainty (see [1]).

5.2.4.5 Self-Cross-Feed

All signal Monte Carlo events, which survive the final selection are considered as signal, regardless if they are reconstructed correctly or not. For instance, a signal event with a decay chain $B \to \tau[\to e\nu\bar{\nu}]\nu$ might be reconstructed as $B \to \tau[\to \pi\nu_\tau]\bar{\nu}_\tau$,

Table 5.16 The self-cross-feed of the reconstructed signal events. The rows contain the decay-channels in which the events where reconstructed. The upper five rows contain the hadronically tagged, and the lower five rows the semileptonically tagged reconstructed decay-channels. The columns indicate the Monte Carlo truth. Each cell contains the fraction of the events in the reconstructed decay-channel (row), which originate from the Monte Carlo decay-channel (column). The label *other* refers to the remaining τ decay-channels, which are not considered in this analysis. The uncertainty due to limited statistics is approximately ± 0.01

	Hadronic						Semileptonic					
	e	μ	π	ρ	a_1	other	e	μ	π	ρ	a_1	other
e	0.91	0.00	0.00	0.00	0.00	0.09	0.00	0.00	0.00	0.00	0.00	0.00
μ	0.00	0.93	0.01	0.02	0.00	0.03	0.00	0.00	0.00	0.00	0.00	0.00
π	0.04	0.19	0.20	0.27	0.04	0.25	0.00	0.00	0.00	0.00	0.00	0.00
ρ	0.01	0.02	0.03	0.68	0.18	0.08	0.00	0.00	0.00	0.00	0.00	0.00
a_1	0.00	0.00	0.00	0.01	0.82	0.17	0.00	0.00	0.00	0.00	0.00	0.00
e	0.05	0.00	0.00	0.00	0.00	0.01	0.86	0.01	0.00	0.00	0.00	0.08
μ	0.00	0.06	0.00	0.00	0.00	0.00	0.01	0.87	0.01	0.02	0.00	0.03
π	0.00	0.01	0.01	0.01	0.00	0.01	0.04	0.17	0.20	0.24	0.04	0.24
ρ	0.00	0.01	0.00	0.04	0.01	0.01	0.01	0.02	0.03	0.64	0.16	0.07
a_1	0.01	0.01	0.00	0.00	0.05	0.01	0.00	0.00	0.00	0.01	0.76	0.15

where the electron was mis-identified as a pion. Hence, the shape of the signal component contains two contributions: correctly reconstructed signal events (that is the signal event is reconstructed in the correct decay channel) and incorrectly reconstructed signal events (that is the signal event is reconstructed in a different decay channel). The last contribution is referred to as **self-cross-feed**.

Table 5.16 states the fraction of the events reconstructed in a decay-channel, which originates from a Monte Carlo decay-channel. There is nearly no self-cross-feed between the hadronic and semileptonic tag. This means, signal events, whose tag-side decayed semileptonically are not reconstructed by the hadronic tag. On the other hand, signal events, whose tag-side decayed hadronically are rarely reconstructed by the semileptonic tag.

The largest self-cross-feed can be found in the $\tau \to \pi\nu_\tau$ decay-channel. Leptons which fail to pass the PID criteria are collected by this channel, as well as $\tau \to \rho\nu_\tau$ decays, where the π^0 is missed. Furthermore, around 25% of the events reconstructed in the pion channel originate from τ decay channels which are not explicitly considered in this analysis. In consequence, the systematic uncertainty of the π channel is potentially large, and is influenced by the τ branching fractions, PID selection efficiencies, and the π^0 reconstruction efficiency.

5.2.4.6 Extra Energy in the Electromagnetic Calorimeter

The sum of all depositions in the electromagnetic calorimeter, which (according to the reconstruction software) do not originate from the reconstructed $\Upsilon(4S)$ candidate, is called the raw extra energy in the electromagnetic calorimeter $E_{\text{ECL}}^{\text{raw}}$.

In order to reduce the dependency on the varying beam-background conditions between the experiments and runs (see Sect. 2.3.3.2), analyses using the extra energy

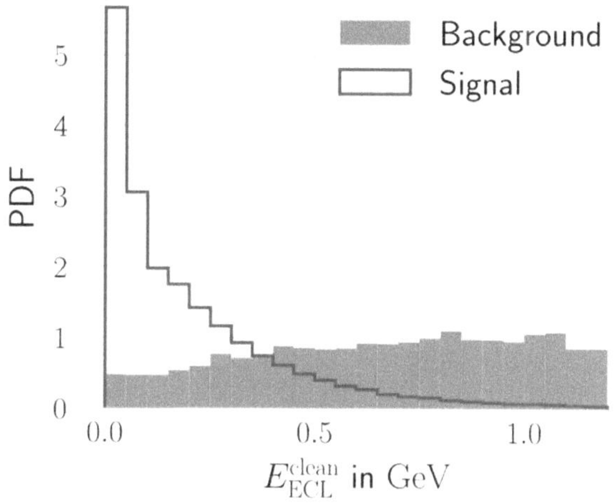

Fig. 5.15 Normed distribution of E_{ECL} for the $\tau \rightarrow e\nu\bar{\nu}$ decay channel with hadronic tag

usually consider only ECL clusters for the calculation of E_{ECL}, which fulfill additional requirements.

In this thesis, only clusters are considered without an associated track and with an energy greater than 50, 100 or 150 MeV if they are located in the barrel, forward end-cap or backward end-cap, respectively. The resulting quantity is the cleaned extra energy in the electromagnetic calorimeter $E_{\mathrm{ECL}}^{\mathrm{clean}}$.

The quantity is shown in Fig. 5.15 for the $\tau \rightarrow e\nu\bar{\nu}$ decay channel with hadronic tag. The signal component peaks at zero, while the background is slowly rising towards higher values. In consequence, $E_{\mathrm{ECL}}^{\mathrm{clean}}$ can be used as a fit variable, to determine the branching fraction of $\mathrm{B} \rightarrow \tau\nu_{\tau}$.

Previous analysis considered more advanced cluster cleanings, e.g. [16] vetoed clusters, which were located close to a reconstructed track if certain criteria based on the ECL cluster shape were fulfilled. The E_{ECL} variable was extensively studied in [17]. It is used in many analyses and its distribution for the signal component was validated on data using double-tagged events [18]. The distribution for background components can be validated on off-resonance data and sidebands (see Sect. 5.3).

For the sake of clarity, I drop the index clean in the following, hence $E_{\mathrm{ECL}} \equiv E_{\mathrm{ECL}}^{\mathrm{clean}}$.

5.2.5 Branching Fraction Extraction

The absolute branching fraction of $\mathrm{B} \rightarrow \tau\nu_{\tau}$ is determined by extracting the number of events containing this decay $N_{\mathrm{sig},d}$ in each decay-channel d with an extended unbinned maximum likelihood fit on the extra energy in the electromagnetic calorimeter E_{ECL} (see Sect. 5.2.4.6)

$$\mathcal{BR}(\mathrm{B} \to \tau\nu_\tau) = \frac{N_{\mathrm{sig},d}}{N_{\mathrm{B\bar{B}}} \cdot \varepsilon_d} \tag{5.20}$$

where $N_{\mathrm{B\bar{B}}} = (7.72 \pm 0.1) \cdot 10^8$ is the number of produced $\mathrm{B\bar{B}}$ events[10] determined by the Belle experiment, and ε_d is the overall reconstruction efficiency in the decay-channel (see Table 5.18).

Figures 5.16 and 5.17 show the distribution of E_{ECL} for all components used in the fit. The continuum component was estimated with a linear regression using the off-resonance sample.

The overall PDF is defined as

$$P(E_{\mathrm{ECL}}) = \sum_{d=\mathrm{e},\mu,\pi,\rho,\mathrm{a}_1} \left(\varepsilon_d N_{\mathrm{B\bar{B}}} \mathcal{BR}(\mathrm{B} \to \tau\nu_\tau) P_{\mathrm{sig},d}(E_{\mathrm{ECL}}) + N_{\mathrm{bkg},d} P_{\mathrm{bkg},d}(E_{\mathrm{ECL}})\right), \tag{5.21}$$

where d denotes the reconstructed decay-channel, and $P_{\mathrm{sig},d}$ and $P_{\mathrm{bkg},d}$ the PDFs obtained from Monte Carlo simulation and off-resonance data for signal and background, respectively. There are six free parameters in the fit: the branching fraction $\mathcal{BR}(\mathrm{B} \to \tau\nu_\tau)$ and the five background normalizations $N_{\mathrm{bkg},d}$. The relative fractions of the different background components was fixed. The fit is performed independently for the hadronic and semileptonic tag.

5.2.5.1 Component Description

The extraction of the branching fraction from E_{ECL} relies on the precise knowledge of the shapes of all background processes. The following background processes were considered during the fit. Each component is described by a histogram PDF using 24 equidistant bins in the region between 0 and 1.2 GeV.

$\Upsilon(4S) \to B^+B^-$ is background from charged $\mathrm{B}^\pm$ decays. Also called generic charged $\Upsilon(4\mathrm{S})$ background. It is the main background component in all decay-channels. In total six (out of ten) streams of Monte Carlo were used to determine the shape of this component.

$\Upsilon(4S) \to B^0\bar{B}^0$ is background from neutral B^0 decays. Also called generic neutral (or mixed) $\Upsilon(4\mathrm{S})$ background. This type of background is suppressed by the tag-side requirements. In total six (out of ten) streams of Monte Carlo were used to determine the shape of this component.

$e^+e^- \to c\bar{c}$ is background from charmed quarks. Also called charm continuum background. Is is the main continuum background in semileptonically tagged decay-channels. This component is estimated from off-resonance data in the final

[10]The Belle analyses assumed equal branching fractions for charged and neutral B mesons. Hence the approximation $N_{\mathrm{B\bar{B}}} \approx 2N_{\mathrm{B^+B^-}}$ is used here. The correction is small compared to the uncertainties of the measurements and therefore neglected.

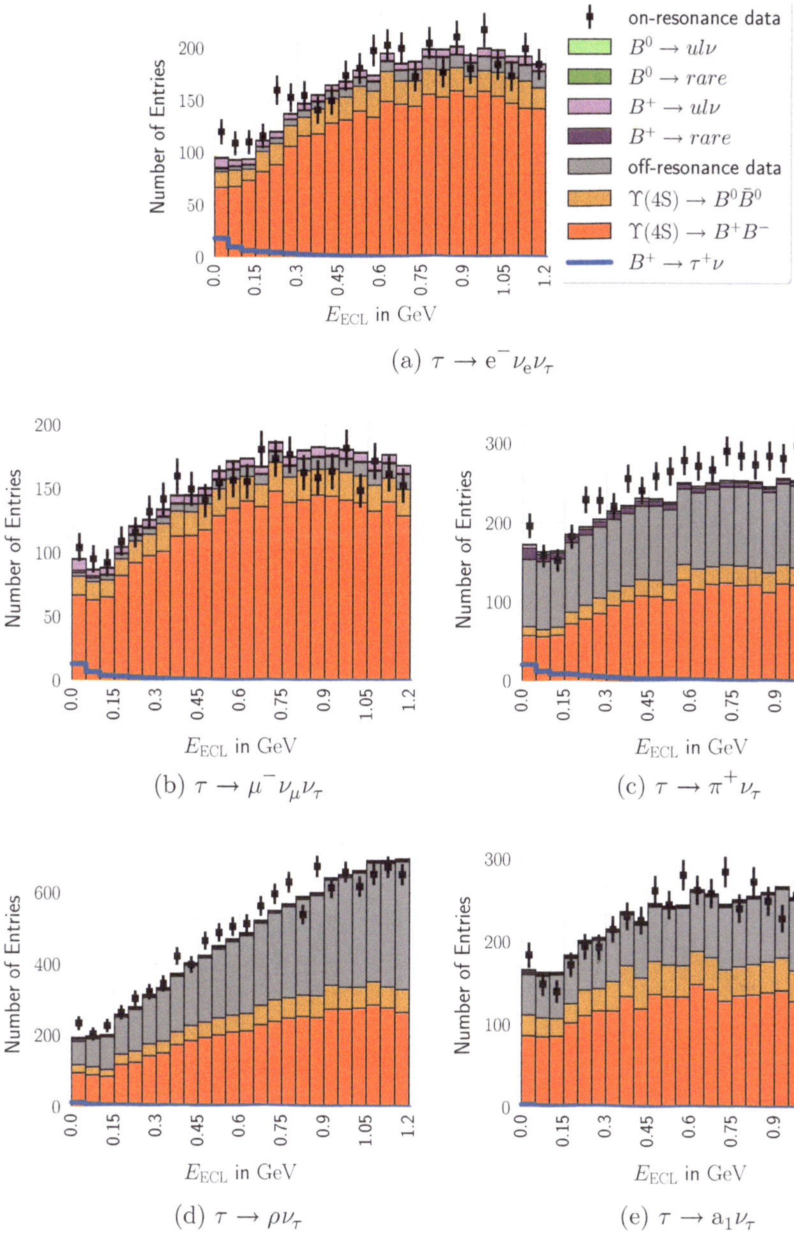

Fig. 5.16 Distribution of E_{ECL} for the hadronically tagged candidates. The stacked histograms show the expectation from Monte Carlo simulation and off-resonance data. The continuum component was estimated with a linear regression using the off-resonance sample

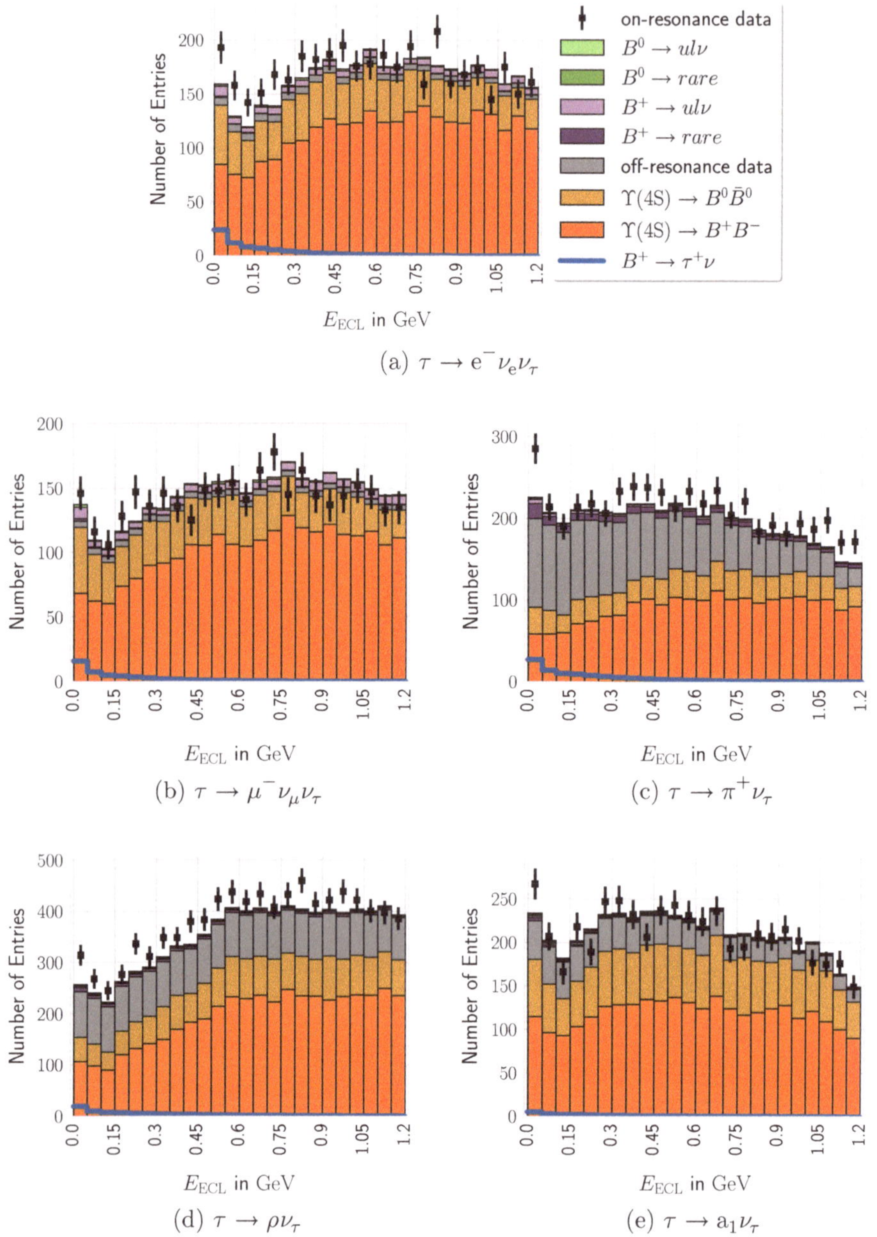

(a) $\tau \to e^-\nu_e\nu_\tau$

(b) $\tau \to \mu^-\nu_\mu\nu_\tau$

(c) $\tau \to \pi^+\nu_\tau$

(d) $\tau \to \rho\nu_\tau$

(e) $\tau \to a_1\nu_\tau$

Fig. 5.17 Distribution of E_{ECL} for the semileptonically tagged candidates. The stacked histograms show the expectation from Monte Carlo simulation and off-resonance data. The continuum component was estimated with a linear regression using the off-resonance sample fit. For the alternative shape description based on the Monte Carlo simulation in total six (out of six) streams of Monte Carlo were used.

$e^+e^- \rightarrow q\bar{q}(uds)$ is background from light quarks. Also called uds continuum background. It is the main continuum background in hadronically tagged decay-channels. This component is estimated from off-resonance data in the final fit. For the alternative shape description based on the Monte Carlo simulation in total six (out of six) streams of Monte Carlo were used.

$B^+ \rightarrow$ rare is background from rare $B^\pm$ decays such as $B^+ \rightarrow \ell^+\nu\gamma$. Although rare decays have an extremely low branching fraction, their final state can be nearly indistinguishable from $B^+ \rightarrow \tau^+\nu_\tau$. In fact, $B^+ \rightarrow \tau^+\nu_\tau$ is contained in this component as well, but was vetoed using Monte Carlo information. In total fifty (out of fifty) streams of Monte Carlo were used to determine the shape of this component.

$B^0 \rightarrow$ rare is background from rare B^0 decays such as $B^0 \rightarrow K\nu\nu$. This type of background is suppressed by the tag-side requirements compared to the charged rare decays. In total fifty (out of fifty) streams of Monte Carlo were used to determine the shape of this component.

$B^+ \rightarrow \bar{u}\ell^+\nu$ is background from CKM-suppressed charged $b \rightarrow u$ transitions. If the decay products from the u are lost during the reconstruction, this component can fake a signal component. In total twenty (out of twenty) streams of Monte Carlo were used to determine the shape of this component.

$B^0 \rightarrow \bar{u}\ell^+\nu$ is background from CKM-suppressed neutral $b \rightarrow u$ transitions. This type of background is suppressed by the tag-side requirements compared to the charged $b \rightarrow u$ transitions. In total twenty (out of twenty) streams of Monte Carlo were used to determine the shape of this component.

$B^+ \rightarrow \tau^+\nu_\tau$ In total ≈ 1200 streams of Monte Carlo were used to determine the shape of this component.

off-resonance data is continuum background recorded 60 MeV below the $\Upsilon(4S)$ resonance. It can be used to estimate the complete continuum background without using Monte Carlo simulation. This includes the above mentioned continuum background from quarks, QED processes and other (even unknown) continuum processes. In total ≈ 0.08 stream of off-resonance data were available and used to determine the shape of the continuum component.

The usually used Belle Monte Carlo campaign did not include QED processes. However, there are Monte Carlo simulated QED events available for Belle, those could not be used during this thesis, due to technical issues. Previous Belle analysis did not use this type of Monte Carlo. Nevertheless, this thesis studied the QED processes using the corresponding Belle II Monte Carlo simulated event for $e^-e^+ \rightarrow \gamma\gamma$, $e^-e^+ \rightarrow e^-e^+$, $e^-e^+ \rightarrow \mu^-\mu^+$, and $e^-e^+ \rightarrow \tau^-\tau^+$. The $\tau^-\tau^+$ component yields a small number of additional continuum events, which are suppressed by the continuum suppression selection (see Fig. 5.14). No candidates from other QED processes were found to pass the selection criteria of this analysis. However, differences between the

Monte Carlo simulation of the continuum component and the off-resonance data are observed (see Sect. 5.3). In consequence, the off-resonance data was used to estimate the continuum background shape.

5.2.5.2 Statistical Uncertainty

The statistical uncertainty of the fit is estimated from the likelihood profile with respect to the branching fraction $\mathcal{BR}(B \to \tau\nu_\tau)$. The significance of the fit is determined by Wilk's theorem (see [16, Sect. 7.7]) using the likelihood ratio of the maximum likelihood obtained from the fit L_{max} and the likelihood under the null-hypothesis L_0 assuming a branching fraction of zero

$$\sigma = \sqrt{2 \ln \frac{L_{\text{max}}}{L_0}}. \tag{5.22}$$

5.3 Validation

Before the final fit on on-resonance data was performed, the distribution of the extra energy in the calorimeter in the different tag-side and signal-side combinations was validated. The background processes were studied using Monte Carlos simulation (Sect. 5.3.1) and data-driven methods (Sects. 5.3.2–5.3.4). The fit procedure was validated (Sect. 5.3.5). And finally possible systematic uncertainties of the analyses were investigated (Sect. 5.3.6).

The following sections summarize the findings of this validation. Tag-side and signal-side combinations which are not explicitly mentioned are in agreement with the expectation.

5.3.1 Monte Carlo Based Study

Monte Carlo simulated events were used to investigate the known background processes. In total six streams were investigated.

5.3.1.1 Leptonic τ Decay Channels

The leptonic τ decay channels were investigated using Monte Carlo simulated events.

The main background originates from semileptonic $B^{\pm}$ decays accompanied by a correctly reconstructed B_{tag} decay: $\Upsilon(4S) \to B_{\text{tag}} B^-[\to D^0 \ell^- \bar{\nu}]$. The additional D^0 meson decays into neutral final state particles like $D^0 \to K_L^0 K_L^0 (K_L^0)$ or $D^0 \to K_L^0 \pi^0$.

The contributing decay channels from neutral $B^0\bar{B}^0$ pairs are very similar $\Upsilon(4S) \to B_{tag}B^0[\to D^-\ell^+\nu]$. The D^- meson decays further into neutral final state particles and an additional charged track, which is assigned to the tag-side.

The $B^+ \to u\ell^+\nu$ component is dominated by the decay $B^+ \to \pi^0\ell^+\nu$, where the π^0 is missed during the reconstruction.

The contribution from rare decays like $B^+ \to \ell^+\nu\gamma$ is negligible small assuming either the branching fractions predicted by the Standard Model or the current experimental limits.

The background from continuum processes described by the Monte Carlo simulation is very small as well. Here the lepton is usually produced via a semileptonic D decay, whereas the tag-side is reconstructed from the remaining charged tracks.

5.3.1.2 Hadronic τ Decay Channels

The hadronic τ decay channels were investigated using Monte Carlo simulated events.

The background from charged B^+B^- pairs is diverse. The main contribution originates from decay-channels with a high branching fraction and accompanied by a correctly reconstructed tag-side. For instance, $\Upsilon(4S) \to B_{tag}B^-[\to D^0\pi^-]$, where the D meson decays further into neutral final state particles like $D^0 \to K_L^0K_L^0(K_L^0)$ or $D^0 \to K_L^0\pi^0$. Another common example is a semileptonic decay-chain $\Upsilon(4S) \to B_{tag}B^-[\to D^0[\to K^-\ell^+\nu]\ell^-\bar{\nu}]$, where the K meson is mis-identified as a pion.

The contributions from neutral $B^0\bar{B}^0$ pairs are even more diverse, since both B mesons are wrongly reconstructed. There is no noticeable pattern, except for the obvious influence of the branching fractions.

The $B^+ \to u\ell^+\nu$ component is dominated by the decay $B^+ \to \pi^0\ell^+\nu$, where the π^0 is missed during the reconstruction and the lepton is mis-identified as a pion. However, due to the additional mis-identification this component is negligible, in contrast to the leptonic decay-channels.

On the other hand, the contribution from rare decays like $B \to K_L^0\pi$ is small.

The background from continuum processes described by the Monte Carlo simulation is as large as the background from $\Upsilon(4S)$ decays. The semileptonically tagged continuum background events are dominated by $e^-e^+ \to c\bar{c}$, because the tag-side requires a lepton, which is rarely produced by light quark pairs: $u\bar{u}$, $d\bar{d}$ and $s\bar{s}$. Hence, light quark pairs are suppressed by the mis-identification rate of leptons.

On the other hand, the hadronically tagged continuum background events have a large contribution from light quarks pairs, because no reconstructed lepton is required. This contribution is as large as the one from $e^-e^+ \to c\bar{c}$. This can be explained by the fact, that the D meson on the tag-side is usually mis-reconstructed, i.e. not matched to a D meson Monte Carlo particle. Probably, true D mesons are suppressed due to their different kinematics compared to the D mesons typically found in $\Upsilon(4S)$ decays. In consequence $c\bar{c}$ pairs and light quark pairs contribute in equal parts to the continuum background in hadronically tagged hadronic τ decay channels.

5.3.2 *Off-Resonance Based Study*

Off-resonance data was used to validate the shape of the continuum component and to identify unknown continuum background processes. The off-resonance was scaled to account for the difference in luminosity recorded on-resonance and off-resonance, and the shifted center-of-mass energy.

5.3.2.1 Peaking Continuum Background in Semileptonically Tagged $\tau \rightarrow e\nu\bar{\nu}$

A large peaking background component in the semileptonically tagged $\tau \rightarrow e\nu\bar{\nu}$ decay channel was observed on off-resonance data by [16]. This peaking background is not described by the Monte Carlo simulation. It was suspected to be caused by $e^-e^+ \rightarrow \gamma[\rightarrow D\bar{D}]\gamma[\rightarrow e^-e^+]$, which is not part of the standard Belle Monte Carlo. A selection on the minimal invariant mass $M_{sig} > 0.2$ GeV of the signal-side electron with all other tracks in the event, and on the minimal invariant mass $M_{tag} > 0.2$ GeV

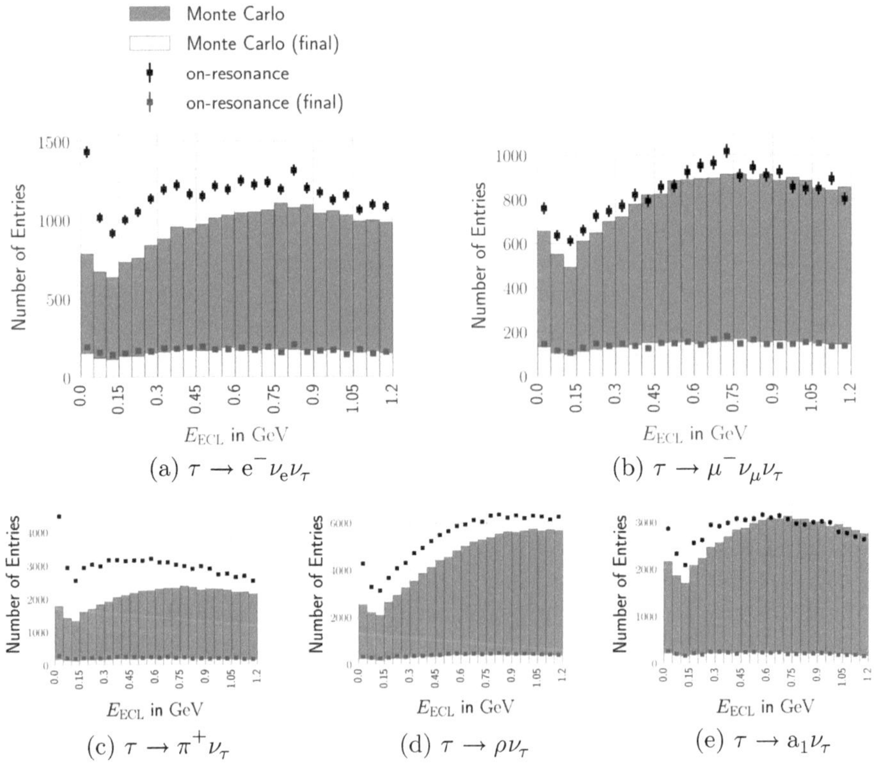

(a) $\tau \rightarrow e^-\nu_e\nu_\tau$ (b) $\tau \rightarrow \mu^-\nu_\mu\nu_\tau$

(c) $\tau \rightarrow \pi^+\nu_\tau$ (d) $\tau \rightarrow \rho\nu_\tau$ (e) $\tau \rightarrow a_1\nu_\tau$

Fig. 5.18 Distribution of E_{ECL} for the semileptonically tagged candidates on on-resonance data. The final selection discards large amounts of the background including its peaking contribution

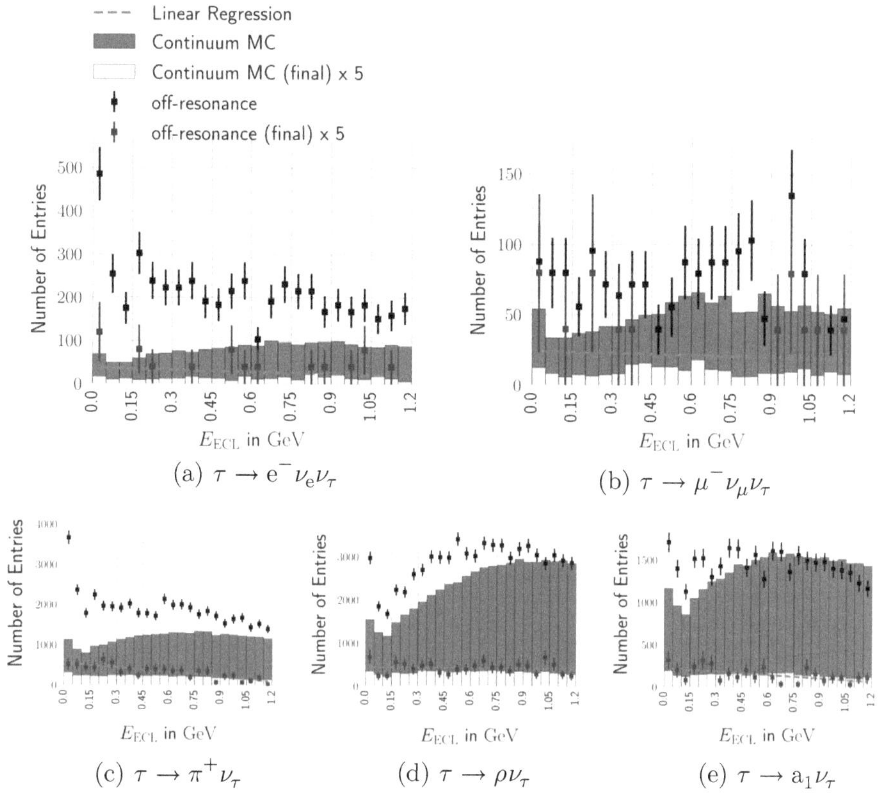

Fig. 5.19 Distribution of E_{ECL} for the semileptonically tagged candidates on off-resonance data. The final selection discards large amounts of the background including its peaking contribution. The shape of the continuum component in the fit is estimated by a linear regression, which is shown as a dashed line

of the tag-side lepton with all other tracks in the event was introduced by [16] to suppress this background. This selection is called invariant-mass-selection in the following.

This analysis observed the same peaking background on on-resonance (see Fig. 5.18) and off-resonance data (see Fig. 5.19). However, the chosen cuts on the `SignalProbability` $\sigma > 0.05$ of the B_{tag} and on the continuum suppression variable $T < 0.85$ (see Sect. 5.2.4.3), discard already all events, which would be discarded by the invariant-mass-selection. This selection is called final-semileptonic-electron-channel-selection in the following.

Figure 5.20 shows the effect of the two selection sets and their relative complement, that is the events which survive the final-semileptonic-electron-channel-selection, but not the invariant-mass-selection. Since the relative complement is negligible small, an additional invariant-mass-selection besides the final-semileptonic-electron-channel-selection does not have any effect. Hence, the invariant-mass-selection was not applied in this thesis

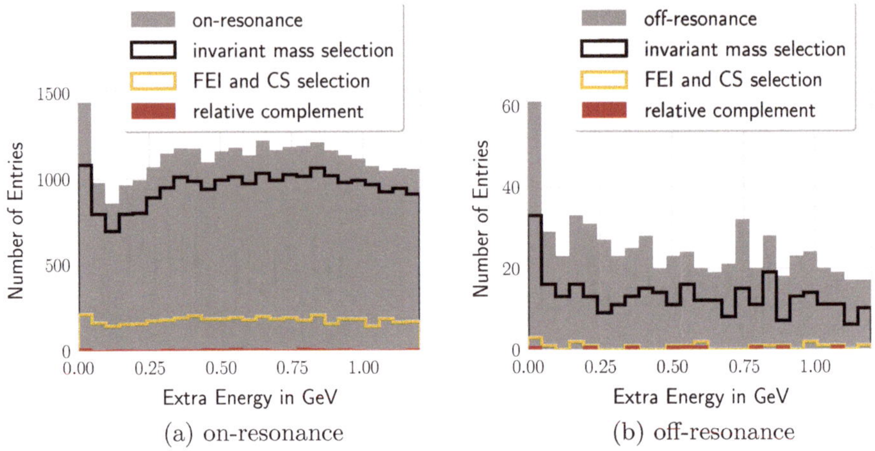

Fig. 5.20 Distribution of E_{ECL} for the $\tau \to e\nu\bar{\nu}$ decay channel. A cut on $M_{\mathrm{sig}} > 0.2\,\mathrm{GeV}$ and $M_{\mathrm{tag}} > 0.2\,\mathrm{GeV}$ is shown in **black** and discards parts of the suspected peaking background. The selection on $\sigma > 0.05$ and $T < 0.85$, used for the semileptonically tagged $\tau \to e\nu\bar{\nu}$ decay channel is shown in yellow and is discards large amounts of the background including its peaking contribution. Finally the relative complement of the invariant mass selection with respect to the σ and T selection is shown in **red**

5.3.2.2 Peaking Continuum Background in Hadronic τ Decay Channels

A large peaking background component in the hadronic τ decay channels is observed on on-resonance and off-resonance data before the final selection on the `SignalProbability` σ and T are applied. This peaking background is not described by the Monte Carlo simulation.

Off-resonance data was used to verify that the final selection $\sigma > 0.01(0.05)$ for hadronic (semileptonic) B_{tag} and $T < 0.7$ discards this peaking background component. From this one can conclude, that the observed peak is caused by an unknown continuum component, and that the component is heavily suppressed by the final selection. The on-resonance and off-resonance distributions of E_{ECL} are shown in Fig. 5.21 and Fig. 5.22, respectively.

Note that this peaking background is different from the one observed in the semileptonically tagged $\tau \to e\nu\bar{\nu}$ decay channel. The invariant-mass-selection introduced in Sect. 5.3.2.1 (where the leptons are replaced by the corresponding hadrons) does not suppress this peaking component.

The previous Belle analysis which used the hadronic tag did not report any observation of this peaking background [1], whereas the previous Belle analysis which used the semileptonic tag did observe severe differences between Monte Carlo simulation and data as well, and obtained the shape for the continuum component solely from off-resonance data. This led to larger systematic uncertainties in the semileptonically tagged measurement compared to the hadronically tagged mea-

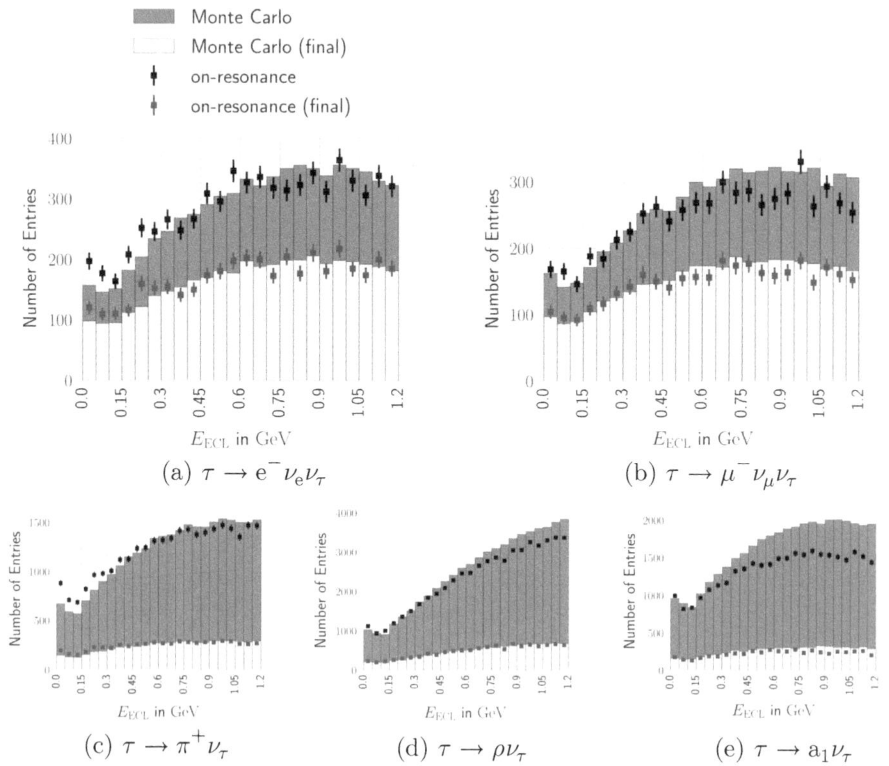

(a) $\tau \to e^{-}\nu_{e}\nu_{\tau}$

(b) $\tau \to \mu^{-}\nu_{\mu}\nu_{\tau}$

(c) $\tau \to \pi^{+}\nu_{\tau}$

(d) $\tau \to \rho\nu_{\tau}$

(e) $\tau \to a_1\nu_{\tau}$

Fig. 5.21 Distribution of E_{ECL} for the hadronically tagged candidates on on-resonance data. The final selection discards large amounts of the background including its peaking contribution

surement, as can be seen in Table 5.17. This thesis estimates the continuum shape from off-resonance as well.

5.3.3 $\mathrm{K_S^0}$ *Sideband Study*

The $\mathrm{K_S^0}$ sideband was used to validate the shape of the overall background and to identify unknown $\Upsilon(4S)$ background processes. The sideband is defined by the presence of two additional tracks forming a good $\mathrm{K_S^0}$ candidate reconstructed by the `FEI`, which is not used for the reconstruction of the $\Upsilon(4S)$ candidate. Subtracting the processes identified using the off-resonance data, this sideband can be used to investigate the shape of $\Upsilon(4S)$ components, in particular the contributions due to $\mathrm{K_L^0}$ mesons.

By the approximate symmetry under the exchange of $\mathrm{K_S^0} \to \mathrm{K_L^0}$, it is expected to detect peaking background due to processes containing $\mathrm{K_L^0}$ in the signal region, where additional $\mathrm{K_S^0}$ candidates are vetoed.

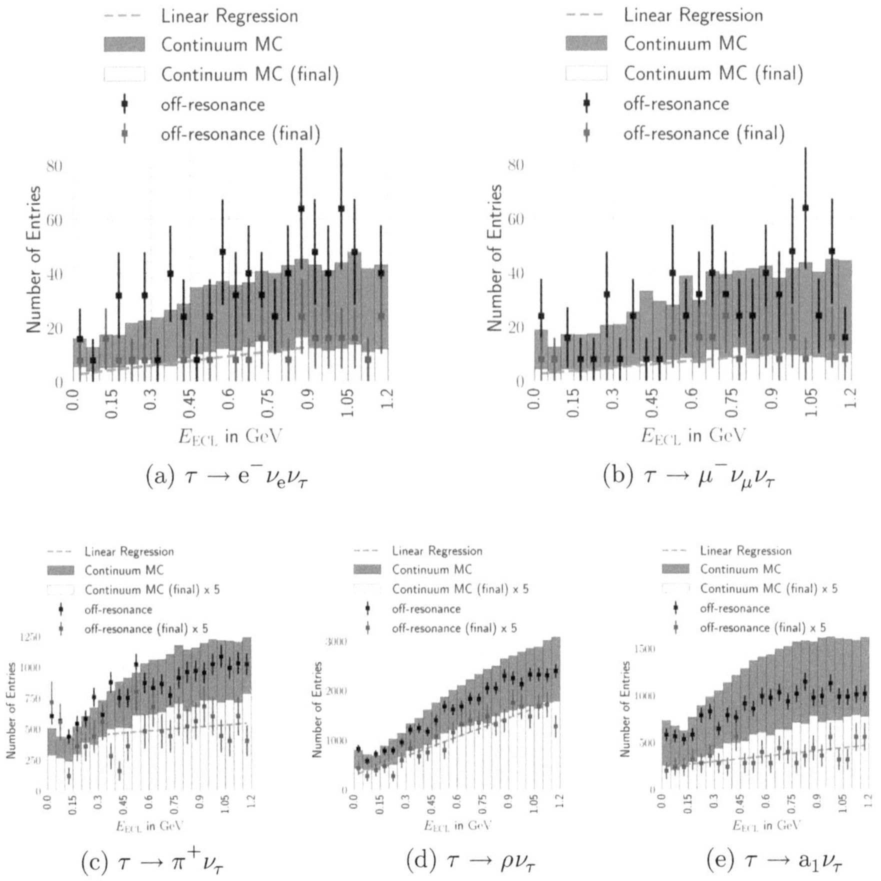

Fig. 5.22 Distribution of E_{ECL} for the hadronically tagged candidates on off-resonance data. The final selection discards large amounts of the background including its peaking contribution. The shape of the continuum component in the fit is estimated by a linear regression, which is shown as a dashed line

Background processes with a K_L^0 can be mitigated by the introduction of a K_L^0 veto, as investigated by [19]. However, K_L^0 are the least understood physics objects at the Belle experiment, hence this leads to a large systematic uncertainty (see Sect. 5.3.6).

There was no significant deviation from the Monte Carlo expectation observed in the hadronically tagged channels.

A possible peaking background component in the semileptonically tagged leptonic τ decay channels was observed after the final selection.

In contrast to the hadronic-tag, the semileptonic-tag used in this thesis was not calibrated. This can cause an incorrect relative fraction between correctly and incorrectly tagged B mesons, and in consequence could explain the observed deviation. The calibration factors for the semileptonic tag-side efficiency are not available, yet. Therefore, this observation was not further investigated during this thesis.

Table 5.17 Relative systematic uncertainties in percent. The previous measurements by Belle based on the hadronic tag [1] and the semileptonic tag [2] are shown in the first and second column respectively. The estimated leading systematic uncertainties of this analysis are shown in the third and fourth column. The tag-side efficiency correction for the semileptonic tag is not yet available, therefore the corresponding uncertainty was estimated with the uncertainty of the hadronic tag-side efficiency correction. The symbol - indicates that the corresponding uncertainty was not investigated

Source	Relative uncertainty in %			
	Ref. [1]	Ref. [2]	Hadronic	Sem.-lep
Tag-side efficiency	7.1	12.6	7.0	≈ 7
PDF shapes	Signal 4.2 Background 8.9	8.5	9.0	6.6
Continuum description	–	14.1	36.3	21.7
τ branching fraction	0.6	0.2	0.4	0.4
Background branching fraction	3.8	3.1	–	–
Tracking efficiency	0.3	0.4	0.4	0.4
π^0 efficiency	0.5	1.1	0.9	1.0
PID efficiency	1.0	0.5	–	–
Signal reconstruction efficiency	0.4	0.6	0.4	0.3
Best candidate selection efficiency	–	0.4	–	–
K_L^0 veto efficiency	7.3	–	–	–
Completeness-constraint efficiency	–	1.9	–	–
Number of $B\bar{B}$ pairs	1.3	1.4	1.3	1.3
Total	14.7	21.2	38.1	23.8

5.3.4 Momentum Shape Study

In addition, the shape of the momentum of the τ in the center-of-mass frame for all candidates in the signal region $E_{ECL} < 100\,\text{MeV}$ was investigated. These distributions are sensitive to unknown peaking contributions from rare decays and $b \rightarrow u\ell\nu$ processes. No unexpected deviations were observed within the statistical fluctuations.

5.3.5 Fitting Procedure

The fitting procedure was validated using six streams of Monte Carlo simulated events. Different options for the smoothing of the PDFs and the constraints used in the fit were investigated. For the final fit it was decided to use the same fitting procedure as in the previous Belle analysis [16]. This ensures a valid comparison.

However, the result of the fit depends heavily on the fit procedure. For instance, using the mis-modeled continuum Monte Carlo simulation, using b-spline based smoothing or adding additional free parameters for the normalization of the continuum component does change the final result and potentially increases the statistical

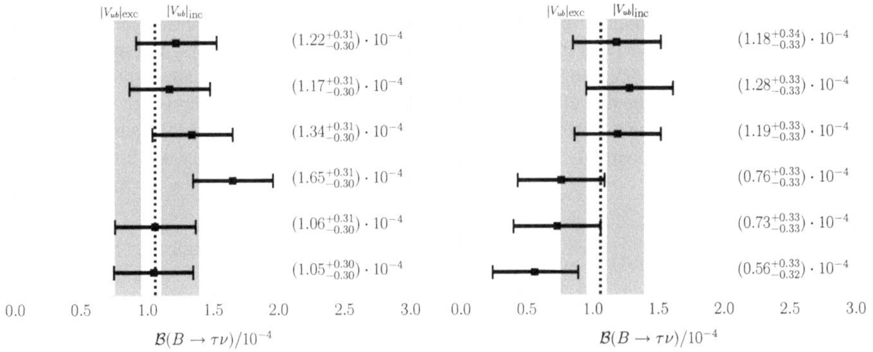

(a) hadronic tag (b) semileptonic tag

Fig. 5.23 Results of the final fit performed on the six streams of Monte Carlo simulated events. The gray bands show the predicted branching fractions by the SM, and the dotted black line the current world average, which was used in the Monte Carlo simulation

uncertainty. To the best knowledge of the author, previous analyses did not take these differences, caused by the fitting procedure, into account. They used either Monte Carlo or off-resonance data, and fixed all relative branching fractions of the background components.

5.3.5.1 Monte Carlo Measurements

The final fit was performed independently on six streams of Monte Carlo simulated events. The tag-side calibration was not used, because it is only applicable to recorded data. In addition, the continuum shape was taken from Monte Carlo for these tests.

As can be seen from Fig. 5.23 the results on Monte Carlo simulation reproduce the simulated branching fraction of $B \rightarrow \tau\nu_\tau$ of $1.06 \cdot 10^{-4}$ within the statistical uncertainties.

5.3.5.2 Continuum Shape Description

Several alternative procedures were investigated to estimate the continuum shape from off-resonance data and Monte Carlo simulation. For the final fit the shape is extracted using a linear regression (i.e. a first-order polynomial was fitted) of the off-resonance data. To estimate the systematic uncertainty, the fit was repeated 100 times by re-sampling the off-resonance data using the bootstrap method (see [20, Sect. 4.3.5]). These fits were performed using first-order, second-order and third-order polynomials. The standard deviation of the fit-results is quoted as the systematic uncertainty due to the continuum description.

The procedure yields a relative uncertainty of 36.3% and 21.7% for the hadronically and semileptonically tagged measurements, respectively. Compared to previous results, this increases the systematic uncertainty of the overall measurements. Previous results did either use the Monte Carlo description [1], or compared the first-order polynomial only with a single second-order polynomial, fitted on the original off-resonance data [2].

5.3.5.3 Closure Tests

The linearity of the fit (see [20, Sect. 10.5]) was tested for 21 different simulated branching fractions between 0 and $5 \cdot 10^{-4}$. For each simulated branching fraction 100 fits were performed using toy Monte Carlo samples. The toy Monte Carlo samples were created by randomly drawing events from the expected signal and background PDFs. On average the fitted branching fraction reproduces the simulated branching fraction, i.e. no bias is observed, as can be seen from Fig. 5.24.

In addition, the pull distribution (see [20, Sect. 10.5]) of the fit was investigated by repeating the final fit using 100 toy Monte Carlo samples. The toy Monte Carlo samples were created by randomly drawing events from the fitted signal and background PDFs.

The obtained pull distributions for the hadronically and semileptonically tagged samples is compatible with a standard normal distribution, as can be seen from Fig. 5.25. In other words: the distribution of the fit is approximately Gaussian, no bias is observed and the calculated statistical uncertainty covers the correct interval.

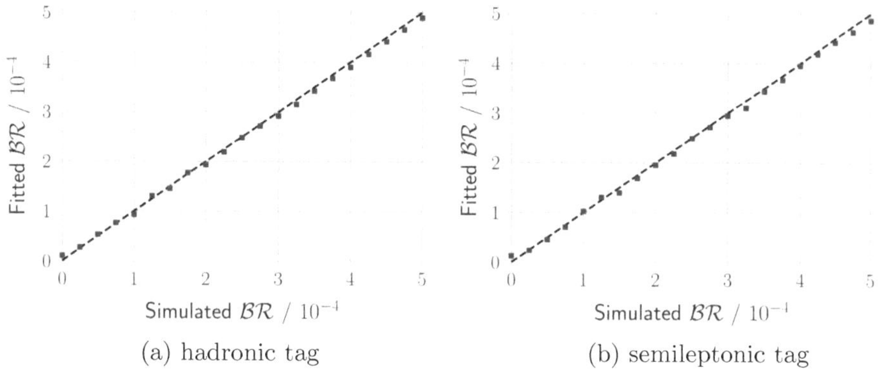

Fig. 5.24 Linearity test of the combined fit for the hadronically and semileptonically tagged sample. The fitted branching fraction $\mathcal{BR}$ is in agreement with the simulated branching fraction over a wide range of possible branching fractions between 0 and $5 \cdot 10^{-4}$

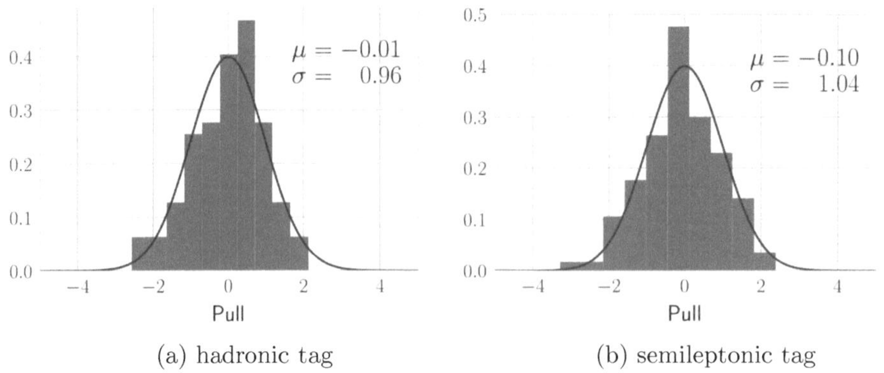

Fig. 5.25 Pull distributions of the combined fit for the hadronically and semileptonically tagged sample. The distribution is in agreement with a standard normal distribution. μ and σ denote the mean and standard deviation of the pull distribution, respectively

5.3.5.4 K_S^0 Sideband

The fit was performed on the K_S^0 sideband using the signal shape of the signal region. As expected, there was no signal observed using the hadronically tagged sample.

A significant excess was observed on the semileptonically tagged sample with a local statistical significance of 2.76 σ. This excess was already discussed in Sect. 5.3.3. Further research is required as soon as the semileptonic tag-side calibration factors for the `FEI` are available.

5.3.6 Systematic Uncertainties

Correctly quantifying the uncertainty of a measured value is one of the most important parts of a scientific measurement. The statistical uncertainty, caused by the finite amount of available data, is theoretically well understood and is either calculated analytically using the likelihood profile of the fit, or is extracted from a Monte Carlo generated distribution.

The systematic uncertainties, caused by deviations from assumptions made by the analyst, are notoriously ill-defined. Usually, in order to estimate the systematic uncertainties, those assumptions are relaxed or alternatives are investigated. Typical sources for systematic uncertainties are the selection efficiencies obtained on Monte Carlo, assumed branching fractions and PDF shapes used in the fit.

The dominating systematic uncertainties in the measurement of $B \to \tau\nu_\tau$ are the uncertainty on the tag-side efficiency correction and the PDF shapes used in the fit, in particular the continuum description. Table 5.17 summarizes all systematic uncertainties investigated by previous Belle measurements and the estimation for this analysis.

This thesis focuses on the dominating systematic uncertainties mentioned above. In a complete analysis it is best practice to quantify all known systematic uncertainties. The known systematic uncertainties and their scaling behavior with increasing luminosity are briefly summarized below.

The tag-side efficiency correction directly enters the calculation of the branching fraction in the denominator of Eq. 5.20. The determination of the hadronic tag-side efficiency and its uncertainty was already discussed in Sect. 4.3.4. The semileptonic tag-side efficiency is not known yet, but is expected to be similar to the hadronic tag-side efficiency, because large parts of the tag-side reconstruction are identical. The associated systematic uncertainty is estimated by combining the statistical and systematic uncertainty of the tag-side efficiency correction factors. This uncertainty decreases with increasing luminosity, because the control channels used in the calibration can be measured with higher precision.

The PDF shapes used in the fit are usually determined from Monte Carlo simulation. The shapes can be sensitive to the statistical fluctuations (discussed here) and mis-modeling of the underlying Monte Carlo sample (see discussion on branching fraction and efficiencies below). The associated systematic uncertainty is estimated by the standard deviation of 100 fits performed using slightly varied PDF shapes, generated by re-sampling the Monte Carlo simulated events using the bootstrap method (see [20, Sect. 4.3.5]). This uncertainty decreases with increasing luminosity, because usually a larger luminosity is accompanied by a larger Monte Carlo sample. Alternatively, PDF shapes can be modeled by (theoretically motivated) analytical functions. Here the difference obtained by choosing different functions can be used to estimate the uncertainty associated with the specific choice used in the fit.

The continuum description determines the shape of the continuum component in the fit. The shape can be either determined by the Monte Carlo simulation (as described above) or taken from the off-resonance data. While the former is poorly modeled (as discussed in Sect. 2.3.3.3), the later suffers from low statistics. In particular the hadronic τ decays are affected by continuum background. In the case of off-resonance data, the systematic uncertainty is estimated by fitting different analytical functions (usually first and second order polynomials) to the off-resonance data. The detailed procedure used in this thesis is described in Sect. 5.3.5.2. The uncertainty decreases with increasing luminosity, because more off-resonance data is available. Furthermore, physical parameters determining the continuum spectra, like branching fractions and fragmentation constants, can be measured more precisely.

The branching fractions of the τ are only known with a finite precision. They enter the calculation of the branching fraction in the denominator of Eq. 5.20. The associated systematic uncertainty is estimated by varying the branching fractions by their known uncertainty. The uncertainty decreases with increasing luminosity, because a B factory like Belle II is as well a τ factory, hence the branching fractions can be measured more precisely. However, the precision is eventually systematically limited.

The background branching fractions which are assumed during the simulation of the Monte Carlo sample, used to determine the PDF shapes, are only known with a finite precision. In particular the branching fractions of peaking background are important, because the induced uncertainty on the normalization of a peaking component cannot be determined by the fit. The associated systematic uncertainty is estimated by varying the weight of the corresponding events and repeating the fit procedure. The uncertainty decreases with increasing luminosity, because the contributing background processes can be measured more precisely. However, the precision of background branching fractions is eventually systematically limited, hence it is expected that at some points the scaling of this uncertainty decouples from the luminosity.

The track reconstruction efficiency has been studied by [21]. It has been found that a systematic uncertainty of 0.35% has to be assigned for each charged track on the signal-side. The uncertainty of the reconstruction efficiency for the tag-side is already accounted for in the tag-side efficiency correction.

The π^0 reconstruction efficiency has been studied by [22]. The deviation between Monte Carlo events and data has been found to be 4%. This uncertainty influences only the $\tau \to \rho\nu_\tau$ decay-channel, hence the associated systematic uncertainty is reduced by the fraction of signal events reconstructed in this channel (see [16]).

The particle identification selection efficiency has been studied by [23, 24]. The deviation between Monte Carlo events and data influences the leptonic τ decays on the signal-side. The uncertainty of the PID selection efficiency for the tag-side is already accounted for in the tag-side efficiency correction.

The signal reconstruction efficiency directly enters the calculation of the branching fraction in the denominator of Eq. 5.20. The statistical uncertainty is usually small, because large amounts of signal Monte Carlo events are used to determine this efficiency. The systematic uncertainty depends on the validity of the τ decay models used in the Monte Carlo generator (see Sect. 5.1.4).

The best-candidate selection can potentially have a different efficiency and modify the shape observed on data. The associated systematic uncertainty can be estimated by choosing a different best-candidate selection criterion, for instance a random best-candidate selection, or by not applying a best-candidate selection at all (as done by [16]).

The K_L^0 veto efficiency has been studied by [19]. It can influences the shape and potentially the selection efficiency for signal and background. Previous Belle analyses (e.g. [1]) used this veto to suppress background processes including K_L^0 particles. The associated systematic uncertainty is estimated by varying the weight of the corresponding events and repeating the fit procedure. Since the KLM clusters used to reconstruct K_L^0 are the least understood physics object of the Belle experiment, the K_L^0 veto was not used during this thesis. In addition the K_L^0 could also influence the E_{ECL} distribution, however this was not studied by previous analyses.

The completeness-constraint efficiency can be studied using double-tagged samples. The fraction of events with additional charged tracks can be compared between Monte Carlo and data. Kronenbitter [16] studied this uncertainty using a double-tagged sample $B^+ \to D^0\pi^+$ and obtained compatible results on Monte Carlo and data with a deviation of 1.9%, which can be used as an estimate for the systematic uncertainty.

The number of $\mathbf{B\bar{B}}$ pairs directly enters the calculation of the branching fraction in the denominator of Eq. 5.20. The Belle collaboration determined $N_{\Upsilon(4S)} = (7.72 \pm 0.1) \cdot 10^8$. Therefore the associated systematic uncertainty is 1.3%. Strictly speaking, this uncertainty is already contained in the tag-side efficiency uncertainty, because the final absolute branching fraction is measured with respect to the investigated control channels, hence the uncertainty already enters in the measurement of the control channels. Nevertheless, the previous Belle measurements added this uncertainty on the number of $B\bar{B}$ pairs to their list of systematic uncertainties.

5.4 Results

The results obtained from the Belle data and a sensitivity estimation for Belle II are summarized in this section.

5.4.1 Belle I

The rare decay channel $B \to \tau\nu_\tau$ was chosen as benchmark, because it was regarded as well-understood and established, however a previously neglected or underestimated systematic uncertainty of the continuum background description was found to be sizeable.

In consequence, the overall uncertainty of the measurement could not profit from the increased statistics, due to the necessary relaxation of the selection criteria and the large systematic uncertainties associated with the peaking backgrounds in the hadronic τ decay modes. These backgrounds are unlikely to be produced by `b2bii` or the `FEI`, since they were neither encountered on the validation of tag-side only (see Sect. 4.3.2) nor in the control channels used for the tag-side calibration (see Sect. 4.3.4). Further research and studies of the encountered background processes are required.

The detailed results of the final fit, including the result of fitting the decay channels independently, are stated in Table 5.18 together with the results of the previous Belle analysis for comparison. The fitted distributions can be found in Figs. 5.26 and 5.27. The combined fits are compatible with the SM and the previous Belle results.

There are two conspicuous individual fits. Firstly, no significant signal was observed in the hadronically tagged $\tau \to \pi\nu_\tau$ channel, which is compatible with the previous individual fit. A more in-depth investigation hints that the linear regression of continuum off-resonance data, is not sufficient to describe the data in the signal

Table 5.18 Results obtained by previous Belle analyses compared to this analysis. The columns show the number of fitted $B \to \tau\nu_\tau$ events N_{sig}, the selection efficiency including the τ branching fractions ε, and the fitted branching fraction $\mathcal{BR}$. The stated selection efficiencies for the hadronic tag include the hadronic tag-side efficiency correction estimated from data. The tag-side efficiency correction for the semileptonic tag was not available at the time of writing, and is therefore **not** included. The combined hadronic tag has a statistical significance of 3.82 σ. The combined semileptonic tag has a statistical significance of 5.20 σ

Decay-channel	Previous			This analysis		
	N_{sig}	$\varepsilon(10^{-4})$	$\mathcal{BR}(10^{-4})$	N_{sig}	$\varepsilon(10^{-4})$	$\mathcal{BR}(10^{-4})$
Hadronic tag [1]						
e^-	16^{+11}_{-9}	3.0	$0.68^{+0.49}_{-0.41}$	116^{+34}_{-33}	6.3	$2.38^{+0.71}_{-0.68}$
μ^-	26^{+15}_{-14}	3.1	$1.06^{+0.63}_{-0.58}$	71^{+32}_{-31}	4.1	$2.21^{+0.99}_{-0.96}$
π	8^{+10}_{-8}	1.8	$0.57^{+0.70}_{-0.59}$	-43^{+64}	9.8	$-0.58^{+0.84}$
ρ	14^{+19}_{-8}	3.4	$0.52^{+0.72}_{-0.62}$	225^{+64}_{-63}	6.2	$4.67^{+1.33}_{-1.31}$
a_1	–	–	–	-5^{+45}	1.6	$-0.39^{+3.54}$
Combined	62^{+23}_{-22}	11.2	$0.72^{+0.27}_{-0.25}$	327^{+89}_{-88}	28.1	$1.51^{+0.41}_{-0.40}$
Semileptonic tag [16]						
e^-	47 ± 25	7.4	0.90 ± 0.47	152^{+40}_{-40}	8.4	$2.34^{+0.62}_{-0.61}$
μ^-	13 ± 21	5.5	0.34 ± 0.55	73^{+36}_{-35}	5.2	$1.81^{+0.90}_{-0.87}$
π	57 ± 21	4.7	1.82 ± 0.68	57^{+60}_{-57}	12.3	$0.60^{+0.63}_{-0.60}$
ρ	119 ± 33	7.3	2.16 ± 0.60	202^{+68}_{-67}	9.6	$2.73^{+0.92}_{-0.90}$
a_1	–	–	–	96^{+63}_{-62}	2.5	$4.97^{+3.24}_{-3.19}$
Combined	222 ± 50	25.0	1.25 ± 0.28	528^{+106}_{-103}	38.0	$1.80^{+0.36}_{-0.35}$

region, and hence the signal contribution is suppressed although there is clearly a signal peak visible (see Fig. 5.26). The large systematic uncertainty due to the continuum description of the hadronically tagged measurements reflects this issue in the final result.

Secondly, the semileptonic $\tau \to a_1\nu_\tau$ channel yields a large branching fraction with large uncertainties. This excess has a local statistical significance of 1.56 σ. This could be caused either by a statistical fluctuation or by the missing tag-side calibration factors for the semileptonic tag.

The achieved relative statistical uncertainty is very similar to the previous analyses. However, in comparison with previous analyses, the analysis presented in this thesis is very conservative.

The final selection was not optimized for maximal statistical significance, instead a phase-space region was chosen which allows an accurate estimation and validation of the background shapes on off-resonance data and sidebands. In particular, the completeness-constraint was not fully exploited, in contrast to previous analyses.

The shape of the continuum component was estimated from off-resonance data and does not rely on Monte Carlo simulation. The previous Belle analysis based on the semileptonic tag took the same approach. Whereas the previous Belle analysis based

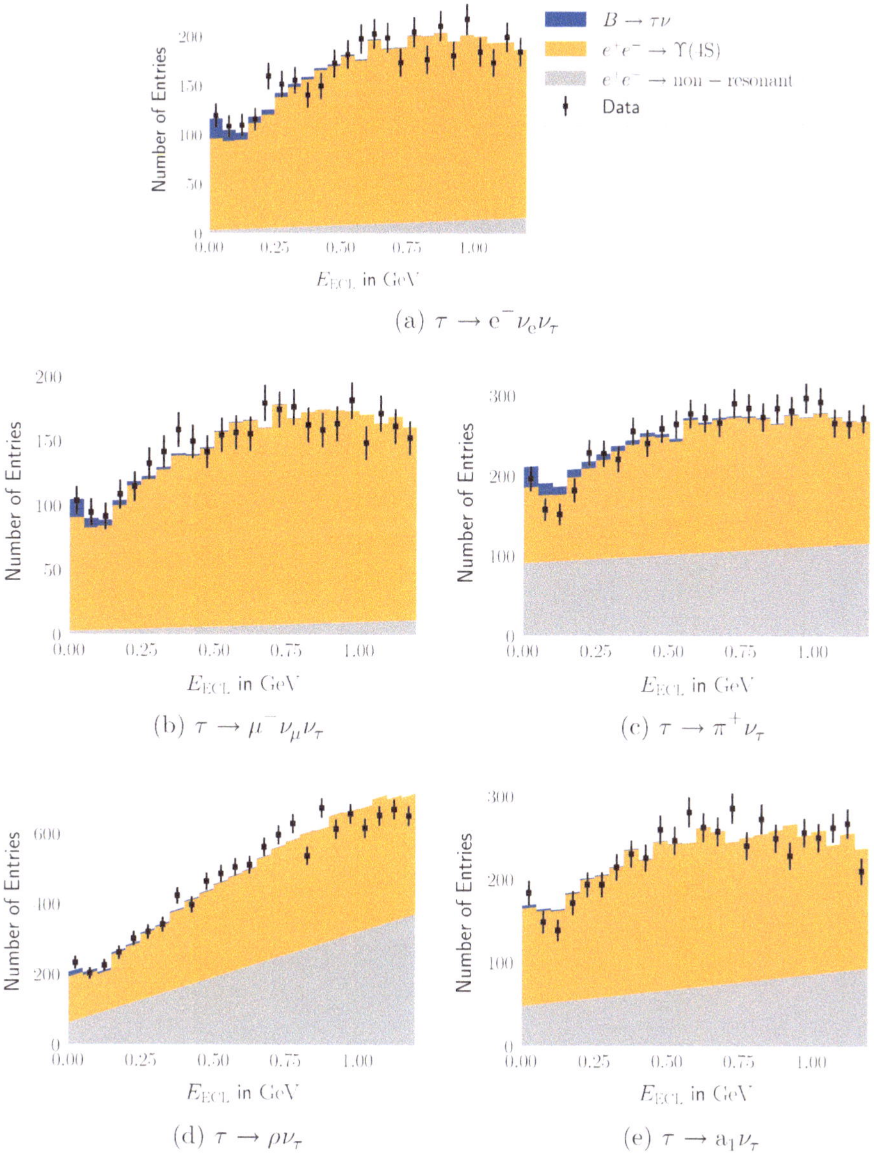

Fig. 5.26 Fitted distribution of E_{ECL} for the hadronically tagged candidates

on the hadronic tag relied on Monte Carlo simulation and consequently reported a smaller systematic uncertainty (see Table 5.17).

This work advocates the use of machine learning. However, to ensure a sound validation of `b2bii` and the `FEI`, only simple cuts on well-understood quanti-

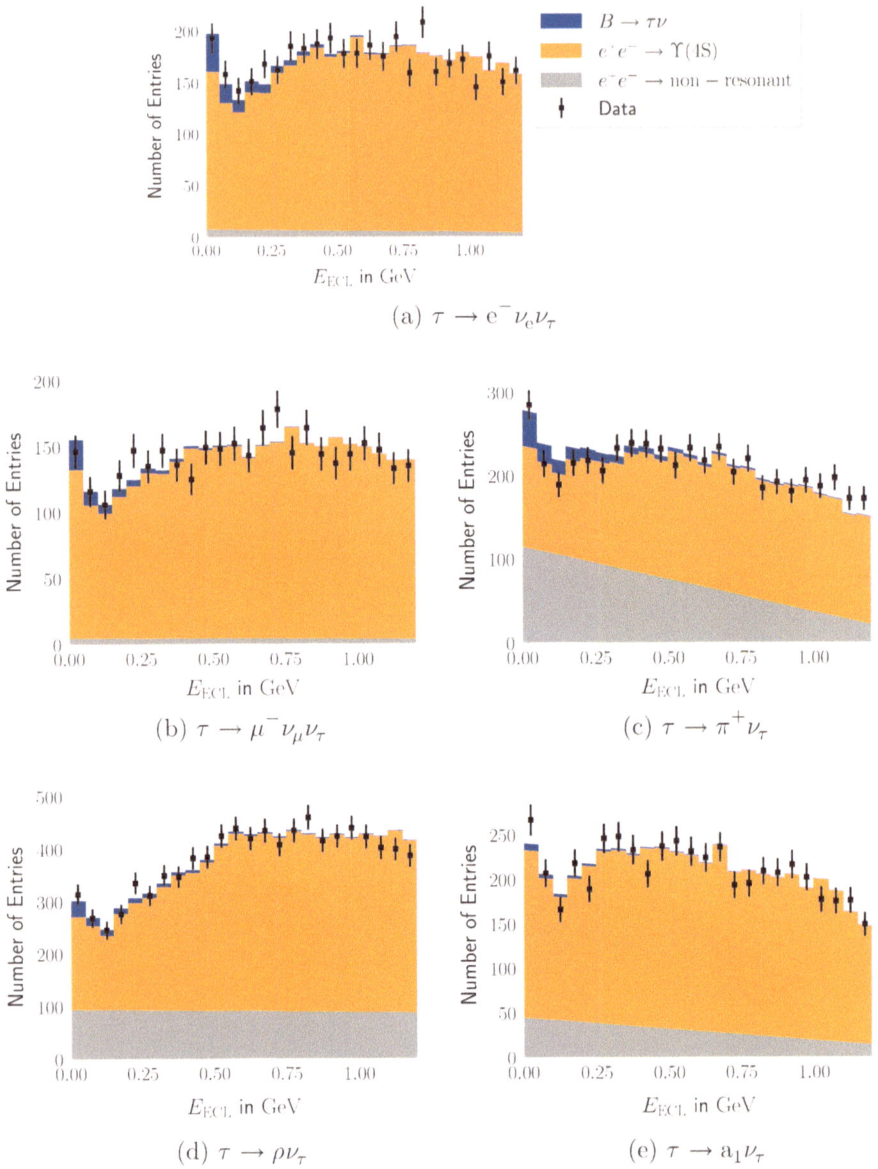

Fig. 5.27 Fitted distribution of E_{ECL} for the semileptonically tagged candidates

ties were used for the signal-side reconstruction and final selection. In particular, this work did not employ a multivariate continuum suppression, which could have introduced additional systematic uncertainties due to the encountered mis-modeling of the continuum processes.

The K_L^0 veto described in [19] was not used, because KLM clusters used to reconstruct K_L^0 are the least understood physics object of the Belle experiment. However, a possible effect on the extra-energy in the electromagnetic calorimeter was studied using the K_S^0 sideband.

The increase in the reconstruction efficiency for the **hadronic tag** is compatible with the expected improvements due to the increased tag-side efficiency. In particular the leptonic τ decay-modes agree well with the expectation. The quoted numbers include the tag-side efficiency correction obtained from control channels on data.

The result for the combined fit using the hadronic tag including statistical and systematic uncertainties is

$$\mathcal{BR}(\mathrm{B} \to \tau\nu_\tau) = \left(1.51^{+0.41}_{-0.40} \pm 0.57\right) \cdot 10^{-4}. \tag{5.23}$$

The shift of the central values with respect to the previous Belle analysis can be explained with the additional uncorrelated systematic uncertainty discovered during this thesis. The previous Belle analysis was too optimistic in the estimation of the systematic uncertainty due to the continuum background. Furthermore, although the same dataset was used, the overlap between the investigated signal events is maximally 40%, due to the relaxed selection criteria and the improved overall signal selection efficiency.

The increase in the expected reconstruction efficiency for the **semileptonic tag** is not as large as expected. One reason is the tight cut on the `SignalProbability` of >0.05, which was necessary due to the encountered peaking backgrounds. Moreover this efficiency does not yet include the tag-side efficiency correction.

The result for the combined fit using the semileptonic tag including statistical and systematic uncertainties is

$$\mathcal{BR}(\mathrm{B} \to \tau\nu_\tau) = \left(1.80^{+0.36}_{-0.35} \pm 0.43\right) \cdot 10^{-4}. \tag{5.24}$$

The shift in the central value with respect to the previous Belle analysis can be explained by the additional systematic uncertainty due to the continuum background. Although the previous analysis used off-resonance data as well to estimate the continuum shape, this thesis relaxed the selection criteria and has a larger overall signal selection efficiency. In consequence, the continuum background description is different. Furthermore, the presented result does not contain the semileptonic tag-side calibration, which was not available at the time of writing. The tag-side calibration influences both: the signal selection efficiency, and the shape of the dominating charged $\mathrm{B\bar{B}}$ background. Finally, although the same dataset was used, the overlap between the investigated events is maximally 65%, due to the relaxed selection criteria and the improved overall signal selection efficiency.

Taking into account the partly uncorrelated statistical uncertainty and the uncorrelated dominating component of the systematic uncertainties, the obtained central values for the branching fraction of $\mathrm{B} \to \tau\nu_\tau$ are compatible with the previous Belle analyses, the previous BaBar analyses, and the SM, as can be seen from Fig. 5.28.

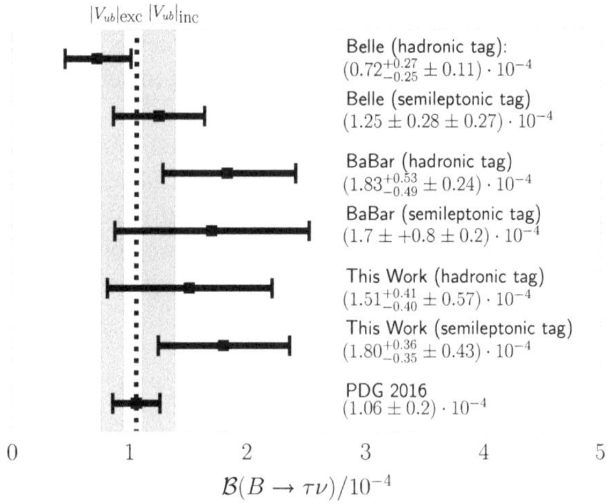

Fig. 5.28 Results obtained by the previous Belle analysis and this thesis. The gray bands show predicted branching fractions by the SM, and the dotted black line the current world average, which was used in the Monte Carlo simulation

5.4.2 *Belle II*

The rare decay B $\rightarrow \tau\nu_\tau$ was used to validate the entire software stack developed during this thesis. The initial goal of validating the software was accomplished. The `b2bii` package, the `mva` package and the `FEI` are ready to be used for physics analyses.

The current version of the reconstruction software of Belle II is not fully optimized yet. During this benchmark analysis an extremely high fake-rate (up to 7 additional tracks in correctly reconstructed events), and a poor ECL resolution with large backgrounds caused by beam-background was observed. A cluster cleaning based on the ECL cluster shapes and cluster timing will be important to mitigate the increased beam-background. However, at the time of writing the necessary reconstruction algorithms are not available in the official Belle II Monte Carlo campaign. It is expected that these **reconstruction issues** will be solved in the future (see also Sect. 4.3.3.2). The current reconstruction issues prevent a meaningful result using the Belle II Monte Carlo simulation.

Assuming that the current reconstruction issues in `BASF2` will be fixed, one can estimate the statistical uncertainty on the branching fraction B $\rightarrow \tau\nu_\tau$ achievable bys Belle II, based on the statistical uncertainty obtained using `b2bii`.

Table 5.19 states the expected relative statistical uncertainty with increasing integrated luminosity. As described in the previous section, the leading systematic uncertainties are expected to scale with the luminosity as well.

Based on the findings described in this thesis, several recommendations for the Belle II experiment can be deduced.

1. The Belle II reconstruction software has to be (and is currently) improved. The fake-rate of tracks has to be decreased significantly.

Table 5.19 Statistical uncertainty: scaled with $\frac{1}{\sqrt{\mathcal{L}}}$ for Belle II

Scenario	Luminosity	Relative uncertainty in %		State
	$\mathcal{L}$ in ab^{-1}	Hadronic	Semileptonic	
Belle	0.711	37.5	22.4	Measured
`b2bii`	0.711	27.1	20.0	Measured
Belle II	1	22.9	16.9	Scaled
Belle II	5	10.2	7.5	Scaled
Belle II	50	3.2	2.4	Scaled

2. The Monte Carlo simulation does not describe the continuum background satisfactorily. A larger off-resonance data sample is paramount to describe the continuum background.
3. The data-driven techniques presented in Sect. 3.3.3 should be used to ensure to suppress background from continuum processes.
4. The extra energy in the electromagnetic calorimeter has to be better understood. Previous analysis primarily tested the shape of the signal component using double-tagged samples [18]. However, the shape of the background distributions (in particular from continuum processes) were not validated to the same degree of accuracy.
5. The tag-side calibration is key to the successful employment of the `FEI`, it should be provided by the collaboration for both: the hadronic and the semileptonic tag.

5.5 Conclusion

The benchmark analysis $B \rightarrow \tau\nu_\tau$ successfully used the entire stack of software developed during this thesis. The `b2bii` package was used to convert the Monte Carlo simulated Belle events and the data recorded by the Belle experiment. The hadronic and semileptonic tag was provided by the `FEI`, which in turn uses the `mva` package.

The expected improvements due to the larger tag-side efficiency were observed in all tag-side and signal-side combinations on Monte Carlo simulated Belle events. The obtained results on recorded data are consistent: with the Monte Carlo expectation, among each other, with the previous Belle analyses and with the Standard Model prediction. The Belle II reconstruction software is not yet mature enough to allow a detailed sensitivity study.

The limitations of the Belle Monte Carlo simulation and current Belle II reconstruction became apparent. The availability of recorded data, provided by the `b2bii` package, proved indispensable. In particular the continuum processes require further research, in order to reduce the systematic uncertainties in measurements which rely on the extra energy in the electromagnetic calorimeter. An increased off-resonance sample size would allow to constrain the shape of the continuum component more precisely.

The analysis presented in this chapter successfully reproduced the previous results reported by Belle and BaBar. It can be used as a template for future analyses which use `b2bii` or the `FEI`.

References

1. K. Hara, Y. Horii, T. Iijima, Evidence for $B^- \to \tau^- \bar{\nu}_\tau$ with a hadronic tagging method using the full data sample of Belle. Phys. Rev. Lett. **110** (2013). https://doi.org/10.1103/PhysRevLett.110.131801
2. B. Kronenbitter et al., Measurement of the branching fraction of $B^+ \to \tau^+ \nu_\tau$ decays with the semileptonic tagging method. Phys. Rev. D **92**(5), 051102 (2015). https://doi.org/10.1103/PhysRevD.92.051102
3. M.D. Schwartz, *Quantum Field Theory and the Standard Model* (Cambridge University Press, 2014). ISBN: 9781107034730
4. A.J. Bevan et al., The physics of the B factories. Eur. Phys. J. C **74**, 3026 (2014). https://doi.org/10.1140/epjc/s10052-014-3026-9
5. C. Patrignani et al., Review of particle physics. Chin. Phys. C **40**(10), 100001 (2016). https://doi.org/10.1088/1674-1137/40/10/100001
6. O. Eberhardt, U. Nierste, M. Wiebusch, Status of the two-Higgs-doublet model of type II. JHEP (2013). https://doi.org/10.1007/JHEP07(2013)118
7. M. Misiak, M. Steinhauser, Weak radiative decays of the B meson and bounds on $M_{H^\pm}$ in the two-Higgs-doublet model. Eur. Phys. J. (2017). https://doi.org/10.1140/epjc/s10052-017-4776-y
8. M. Ziegler, Search for the decay $B \to \tau^- \tau^+$ with the belle experiment. Ph.D. thesis, KIT (2016). https://doi.org/10.5445/IR/1000063630
9. B. Aubert et al., Search for the rare decay $B^0 \to \tau^+ \tau^-$ at BABAR. Phys. Rev. Lett. **96** (2006). https://doi.org/10.1103/PhysRevLett.96.241802
10. R. Aaij et al., Search for the decays $B_s^0 \to \tau^+ \tau^-$ and $B^0 \to \tau^+ \tau^-$. Phys. Rev. Lett. **118**(25) (2017). https://doi.org/10.1103/PhysRevLett.118.251802
11. J.H. Kühn, A. Santamaria, Tau decays to pions. Z. Phys. (1990). https://doi.org/10.1007/BF01572024
12. J.P. Lees et al., Evidence of $B^+ \to \tau^+ \nu$ decays with hadronic B tags. Phys. Rev. D **88**, 031102 (2013). https://doi.org/10.1103/PhysRevD.88.031102
13. B. Aubert et al., Search for $B^+ \to \ell^+ \nu_\ell$ recoiling against $B^- \to D^0 \ell^- \bar{\nu} X$. Phys. Rev. D **81**, 051101 (2010). https://doi.org/10.1103/PhysRevD.81.051101
14. J. Schwab, Calibration of the full event interpretation for the Belle and the Belle II experiment. MA thesis, KIT (2017), https://ekp-invenio.physik.uni-karlsruhe.de/record/48931
15. E.I. Shibata, Hadronic structure in $\tau^- \to \pi^- \pi^- \pi^+ \nu_\tau$ decays. Tau Lepton Phys. Proc. C 0209101 (2002), TU05. https://doi.org/10.1016/S0920-5632(03)80305-5
16. B. Kronenbitter, Measurement of the branching fraction of $B^+ \to \tau^+ \nu_\tau$ decays at the Belle experiment. Ph.D. thesis, KIT (2014), https://ekp-invenio.physik.uni-karlsruhe.de/record/48604
17. J. Grygier, Development and evaluation of a new calorimeter based variable for missing energy analyses at the Belle experiment. MA thesis, KIT (2013), https://ekp-invenio.physik.uni-karlsruhe.de/record/48043
18. R. Barlow, Experimental status of $Bto\tau\nu$ and $Btol\nu(\gamma)$, in *Proceedings of the 6th International Workshop on the CKM Unitarity Triangle, CKM 2010*, Warwick, UK, 6–10 September 2010 (2011). arXiv: 1102.1267

19. K. Hara, Calibration of low momentum K_L^0 efficiency for veto usage in missing E analyses. Belle Note 1228 (internal)
20. O. Behnke, K. Kroeninger, T. Schoerner-Sadenius, G. Schott, *Data Analysis in High Energy Physics* (Wiley-VCH, 2013). ISBN: 9783527410583
21. B. Bhuyan, High P_T tracking efficiency using partially reconstructed D^* decays. Belle Note 1165 (internal)
22. S.W. Lin, π^0 systematics using inclusive η. Belle Note 645 (internal)
23. S. Nishida, Study of Kaon and Pion identification using inclusive D^* sample. Belle Note 779 (internal)
24. L. Hinz, Lepton ID efficiency correction and systematic error. Belle Note 954 (internal)

Chapter 6
Conclusion

This Ph.D. thesis covered four major topics.

The Belle to Belle II Conversion package (`b2bii`) enables Belle II physicists to analyze the dataset recorded by Belle using `BASF2`. Thus, the entire Belle II analysis software stack was validated on recorded data, years before recorded data from the Belle II experiment was available.

The multivariate analysis package (`mva`) enables Belle II physicists to keep up with the rapid developments in the field and to easily employ modern machine learning algorithms in their work. Most of the multivariate methods used in the reconstruction and analysis algorithms in `BASF2` are built on the `mva` package and use the default classification method `FastBDT`, both developed during this thesis.

The Full Event Interpretation algorithm (`FEI`) enables Belle II physicists to measure a wide range of interesting decays with a minimum amount of detectable information. The `FEI` more than doubles the tag-side efficiency compared to its (already very successful) predecessor.

The B $\rightarrow \tau\nu$ benchmark analysis validated the entire Belle II analysis software stack. Unresolved reconstruction issues in the Belle II framework and previously unknown background contributions were discovered before Belle II started recording data. It successfully reproduced the previous results reported by Belle. The analysis serves as a prototype for other current (using `b2bii`) and upcoming exclusively tagged analyses.

All software packages developed during this thesis were validated on data and are used in production by Belle II physicists all over the world.

T. Keck, *Machine Learning at the Belle II Experiment*, Springer Theses,
https://doi.org/10.1007/978-3-319-98249-6_6

Appendix A
B2BII

A.1 Monitoring Histograms

The full list of extracted quantities used for validating the Belle to Belle II Conversion. The number in parenthesis states the number of quantities.

Beam Parameters (16):

- the experiment, run and event number (3);
- the energy in the HER (High Energy Ring), LER (Low Energy Ring) and CMS (Center Of Mass System) (3);
- the crossing angle between the HER and LER (1);
- and the position and uncertainty of the interaction point (9).

K_S^0, Λ, **and converted** γ **(201)**:

- invariant mass, four-momentum and vertex position (8);
- four-momentum and POCAs (point of closest approach) of the daughters (14);
- PID information calculated by the K_S^0 finder (4);
- invariant mass, four-momentum and vertex position after a mass-constraint vertex fit (8);
- uncertainties of the four-momenta and vertex position calculated by the mass-constrained vertex-fit (28);
- p-value of the mass-constrained vertex fit (1);
- and Monte Carlo truth of the total momentum and PDG value of the daughters (4);

K_L^0 **(9)**:

- KLM cluster position (3);
- number of KLM layers with hits (1);
- innermost KLM layer with hits (1);
- and Monte Carlo truth of the four-momentum and PDG value (4);

T. Keck, *Machine Learning at the Belle II Experiment*, Springer Theses,
https://doi.org/10.1007/978-3-319-98249-6

Tracks (41):

- PID information including quality indicators (6);
- four-momentum and POCA (7);
- uncertainties of the four-momentum and POCA (28);

γ **(48)**:

- four-momentum and ECL cluster position (7);
- uncertainties of the four-momentum and cluster position (28);
- spherical coordinates of the ECL cluster position (3);
- ECL cluster shape quantities (6);
- and Monte Carlo truth of the four-momentum (4);

π^0 **(41)**:

- four-momentum and vertex-position (7);
- invariant mass calculated using four-momentum and the daughter four-momenta (2)
- uncertainties of the four-momentum and vertex position (28);
- and Monte Carlo truth of the four-momentum (4);

MC Information (12):

- four-momentum of the MC particle (4);
- PDG code of the MC particle (1);
- vertex position or POCA of the MC particle (3);
- PDG code of the mother of charged and neutral pions (3);
- and number of daughters (1)

Appendix B
FEI

B.1 Default Configuration

A detailed description of the current default configuration of the `Full Event Interpretation` is given in this section. In the default configuration all employed multivariate classifiers have the same configuration: a boosted decision tree (`FastBDT`) with 400 trees, a depth of 3 per tree, a shrinkage of 0.1 and a sub-sampling rate (or bagging rate) of 0.5. The output of the boosted decision tree is named σ below. There is no cut applied on the output of the vertex fitting, however the information of the vertex fitting is used in the multivariate classifier.

All candidates reconstructed from one specific decay-channel are ranked according to a variable and only the n best candidates are kept for each decay-channel—this is known as **pre-cut**. A multivariate classifier is applied to all candidates. Afterwards, all candidates reconstructed from a specific particle are ranked according to the output of the associated multivariate classifier and only the m best candidates are kept for each particle—this is known as **post-cut**. The pre-cut and post-cut of all particles is summarized in Table B.1.

In the remainder of this chapter, the particles explicitly reconstructed by the `FEI` are discussed in detail. The `FEI` does not explicitly reconstruct intermediate resonances like ρ, ω or K^*, however these resonances are implicitly exploited by the multivariate classifiers applied to all candidates. The explicitly reconstructed decay-channels are listed below. The **highlighted** decay-channels were already used by the `Full Reconstruction (FR)` algorithm employed by Belle. The charge-conjugated particle and decay-channels are always implied throughout the text.

B.1.1 Final State Particles

Particles which can directly be reconstructed from `Tracks`, `ECL clusters`, `KLM clusters`, or `V0 objects` are considered final state particles by the `FEI`.

T. Keck, *Machine Learning at the Belle II Experiment*, Springer Theses,
https://doi.org/10.1007/978-3-319-98249-6

Table B.1 Pre-cut: ranking criteria and number of best candidates which is kept for each decay-channel before the vertex fitting and the multivariate classifier application. Post-cut: ranking criteria and number of best candidates which is kept in total after the multivariate classifier was applied. P-ID is the particle identification information of particle P from the PID sub-detectors, E is the energy of the particle, σ is the signal probability of the particle outputted by the corresponding multivariate classifier, M is the invariant mass, M_{P} is the nominal mass of particle P, Q is the released energy in the decay, Q_{P} is the nominal released energy in the decay of the particle P, and $\prod_i \sigma_i$ is the product over the signal probabilities of all daughter particles

Particle	Pre-cut	Post-cut
e^+	10 highest e-ID	5 highest σ and $0.01 < \sigma$
μ^-	10 highest μ-ID	5 highest σ and $0.01 < \sigma$
π^+	20 highest π-ID	10 highest σ and $0.01 < \sigma$
K^+	20 highest K-ID	10 highest σ and $0.01 < \sigma$
γ	40 highest E	20 highest σ and $0.01 < \sigma$
π^0	20 lowest $\|M - M_{\pi^0}\|$	10 highest σ and $0.01 < \sigma$
K_S^0	20 lowest $\|M - M_{K_S^0}\|$	10 highest σ and $0.01 < \sigma$
K_L^0	20 lowest $\|M - M_{K_L^0}\|$	10 highest σ and $0.01 < \sigma$
D^0 (had)	20 lowest $\|M - M_{D^0}\|$	10 highest σ and $0.001 < \sigma$
D^0 (sem)	20 highest $\prod_i \sigma_i$	10 highest σ and $0.001 < \sigma$
D^0 (klong)	20 highest $\prod_i \sigma_i$	10 highest σ and $0.001 < \sigma$
D^+ (had)	20 lowest $\|M - M_{D^+}\|$	10 highest σ and $0.001 < \sigma$
D^+ (sem)	20 highest $\prod_i \sigma_i$	10 highest σ and $0.001 < \sigma$
D^+ (klong)	20 highest $\prod_i \sigma_i$	10 highest σ and $0.001 < \sigma$
D^{+*} (had)	20 lowest $\|Q - Q_{D^{+*}}\|$	10 highest σ and $0.001 < \sigma$
D^{+*} (sem)	20 lowest $\|Q - Q_{D^{+*}}\|$	10 highest σ and $0.001 < \sigma$
D^{+*} (klong)	20 lowest $\|Q - Q_{D^{+*}}\|$	10 highest σ and $0.001 < \sigma$
D_S^+ (had)	20 lowest $\|M - M_{D_S^+}\|$	10 highest σ and $0.001 < \sigma$
D_S^+ (klong)	20 highest $\prod_i \sigma_i$	10 highest σ and $0.001 < \sigma$
D_S^{+*} (had)	20 lowest $\|Q - Q_{D_S^{+*}}\|$	10 highest σ and $0.001 < \sigma$
D_S^{+*} (klong)	20 lowest $\|Q - Q_{D_S^{+*}}\|$	10 highest σ and $0.001 < \sigma$
B^+ (had)	20 highest $\prod_i \sigma_i$	20 highest σ
B^+ (sem)	20 highest $\prod_i \sigma_i$	20 highest σ
B^+ (klong)	20 highest $\prod_i \sigma_i$	20 highest σ
B^0 (had)	20 highest $\prod_i \sigma_i$	20 highest σ
B^0 (sem)	20 highest $\prod_i \sigma_i$	20 highest σ
B^0 (klong)	20 highest $\prod_i \sigma_i$	20 highest σ

B.1.1.1 Charged

The charged final state particles e^+, μ^+, π^+ and K^+ are reconstructed from `Tracks` which are close to the IP, that is a maximum radial distance of 2 cm and maximum distance along the beam-pipe of 4 cm).

Only particles from the primary interaction are regarded as signal. The input features for the multivariate classifiers include: the likelihoods of the PID detectors; kinematics and tracking variables; the position in the best-candidate ranking of the pre-cut.

Protons are not used in the default configuration of the `FEI`.

B.1.1.2 Neutral

Photons γ are reconstructed from `ECL clusters` and `V0 objects`. Only photons with an energy of at least 140, MeV (100 MeV) in the forward, 130 MeV (50 MeV) in the barrel, and 200 MeV (150 MeV) in the backward region, are considered. The numbers in parenthesis are for converted Belle data.

Only particles from the primary interaction are regarded as signal. The input features for the multivariate classifiers include: kinematic variables and the position in the best-candidate ranking of the pre-cut. Additionally, cluster information like the numbers of hits and the timing is used for photons reconstructed from `ECL clusters`.

Λ mesons (reconstructed from `V0 objects`) and neutrons are not used in the default configuration of the `FEI`. K^0_S and K^0_L mesons are discussed below.

Neutrinos ν are not explicitly reconstructed by the `FEI`, because they exit the Belle and Belle II detector undetected. Therefore, the `FEI` treats decay-channels containing a neutrino separately from other decay-channels, due vastly different uncertainties on the kinematics.

B.1.2 Light Neutral Mesons

B.1.2.1 π^0 Neutral Pion

Neutral pions π^0 are reconstructed from two photons.

1. $\pi^0 \to \gamma\gamma$

For Belle II the `FEI` does this reconstruction itself, on converted Belle data the **default π^0 reconstruction** of the `B2BII` package is used. Only π^0 candidates with an invariant mass between 80 and 180 MeV are considered.

The input features for the multivariate classifiers include: kinematic variables and the position in the best-candidate ranking of the pre-cut. In addition the signal probability σ of the two daughter photons is used for Belle II.

B.1.2.2 K_S^0 Neutral Short-Lived Kaon

Neutral kaons K_S^0 are reconstructed from two charged pions, two neutral pions or a `V0 object`.

$$1.\ K_S^0 \to \pi^+\pi^- \qquad 2.\ K_S^0 \to \pi^0\pi^0$$

For converted Belle data only the `V0 objects` are used. Only K_S^0 candidates with an invariant mass between 400 and 600 MeV are considered.

The input features for the multivariate classifiers include: kinematic and vertex position variables, and the position in the best-candidate ranking of the pre-cut. In addition the information provided from the Belle K_S^0 finder is used for converted Belle data. For Belle II the signal probability σ, tracking variables and the total momentum in the rest frame of the K_S^0, of each daughter is used.

B.1.2.3 K_L^0 Neutral Long-Lived Kaon

Neutral kaons K_L^0 are reconstructed from `KLM clusters`. All candidates are considered. The usual signature of hits in the KLM without an associated track does only provide a very rough estimate of the direction and energy of the assumed K_L^0. Therefore, the `FEI` treats decay-channels containing K_L^0 particles separately from other decay-channels, due to the vastly different uncertainties on the kinematics.

The input features for the multivariate classifiers include: the energy and the `KLM cluster` timing information.

B.1.3 Charmed Mesons

All charmed mesons have the same input features for the multivariate classifiers: lorentz invariant kinematic and vertex variables of the charmed meson; kinematic, tracking and vertex variables of all daughters evaluated in the rest frame of the charmed meson; the decay mode identifier and signal probability σ of all daughters; the invariant mass of all possible combinations of daughter particles; and the position of the candidate in the best candidate ranking of the pre-cut.

Since all kinematic and vertex informations are either Lorentz-invariant like the invariant mass or evaluated in the rest frame of the charmed meson, the performance of the multivariate classifier is independent of the B meson decay from which the charmed meson originated.

B.1.3.1 D^0 Neutral D Meson from Hadronic Decay-Channels

The following hadronic channels are used to reconstruct D^0 mesons:

1. $D^0 \to K^-\pi^+$
2. $D^0 \to K^-\pi^+\pi^0$
3. $D^0 \to K^-\pi^+\pi^0\pi^0$
4. $D^0 \to K^-\pi^+\pi^+\pi^-$
5. $D^0 \to K^-\pi^+\pi^+\pi^-\pi^0$
6. $D^0 \to \pi^-\pi^+$
7. $D^0 \to \pi^-\pi^+\pi^0$
8. $D^0 \to \pi^-\pi^+\pi^0\pi^0$
9. $D^0 \to \pi^-\pi^+\pi^+\pi^-$
10. $D^0 \to K_S^0\pi^0$
11. $D^0 \to K_S^0\pi^+\pi^-$
12. $D^0 \to K_S^0\pi^+\pi^-\pi^0$
13. $D^0 \to K^-K^+$
14. $D^0 \to K^-K^+\pi^0$
15. $D^0 \to K^-K^+K_S^0$

Only hadronic D^0 candidates with an invariant mass between 1.7 and 1.95 GeV are considered. If not stated otherwise D^0 always refers to a D^0, which decayed in one of the hadronic decay modes listed above, throughout the text.

B.1.3.2 D^0 Neutral D Meson from Semileptonic Decay-Channels

Due to the different kinematic properties, D^0 mesons from semileptonic decay-channels are processed separately by the FEI. The following semileptonic channels are used to reconstruct D^0 mesons:

1. $D^0 \to K^-e^+\nu$
2. $D^0 \to K^-\mu^+\nu$
3. $D^0 \to K^-\pi^0e^+\nu$
4. $D^0 \to K^-\pi^0\mu^+\nu$
5. $D^0 \to K_S^0\pi^-e^+\nu$
6. $D^0 \to K_S^0\pi^-\mu^+\nu$

There are no further cuts applied to semileptonic D^0 candidates.

B.1.3.3 D^0 Neutral D Meson from Decay-Channels with K_L^0 Neutral Long-Lived kaons

Due to the different kinematic properties, D^0 mesons from decay-channels which include a K_L^0 are processed separately by the FEI. The following channels are used to reconstruct D^0 mesons:

1. $D^0 \to K_L^0\pi^0$
2. $D^0 \to K_L^0\pi^+\pi^-$
3. $D^0 \to K_L^0\pi^+\pi^-\pi^0$
4. $D^0 \to K^-K^+K_L^0$

There are no further cuts applied to D^0 candidates with K_L^0.

B.1.3.4 D^+ Charged D Meson from Hadronic Decay-Channels

The following hadronic channels are used to reconstruct D^+ mesons:

1. $D^+ \to K^-\pi^+\pi^+$
2. $D^+ \to K^-\pi^+\pi^+\pi^0$
3. $D^+ \to K^-K^+\pi^+$
4. $D^0 \to K^-K^+\pi^+\pi^0$
5. $D^+ \to \pi^+\pi^0$
6. $D^+ \to \pi^+\pi^+\pi^-$
7. $D^+ \to \pi^+\pi^+\pi^-\pi^0$
8. $D^+ \to K_S^0\pi^+$
9. $D^+ \to K_S^0\pi^+\pi^0$
10. $D^+ \to K_S^0\pi^+\pi^+\pi^-$
11. $D^+ \to K^+K_S^0K_S^0$

Only hadronic D^+ candidates with an invariant mass between 1.7 and 1.95 GeV are considered. If not stated otherwise D^+ always refers to a D^+, which decayed in one of the hadronic decay modes listed above, throughout the text.

B.1.3.5 D^+ Charged D Meson from Semileptonic Decay-Channels

Due to the different kinematic properties, D^+ mesons from semileptonic decay-channels are processed separately by the FEI. The following semileptonic channels are used to reconstruct D^+ mesons:

1. $D^+ \to K^0_S e^+ \nu$
2. $D^+ \to K^0_S \mu^+ \nu$
3. $D^+ \to K^- \pi^+ e^+ \nu$
4. $D^+ \to K^- \pi^+ \mu^+ \nu$

There are no further cuts applied to semileptonic D^+ candidates.

B.1.3.6 D^+ Charged D Meson from Decay-Channels with K^0_L Neutral Long-Lived kaons

Due to the different kinematic properties, D^+ mesons from decay-channels which include a K^0_L are processed separately by the FEI. The following channels are used to reconstruct D^+ mesons:

1. $D^+ \to K^0_L \pi^+$
2. $D^+ \to K^0_L \pi^+ \pi^0$
3. $D^+ \to K^0_L \pi^+ \pi^+ \pi^-$
4. $D^+ \to K^+ K^0_L K^0_S$
5. $D^+ \to K^+ K^0_L K^0_L$

There are no further cuts applied to D^+ candidates with K^0_L.

B.1.3.7 J/ψ Meson J/psi Resonance

The following hadronic channels are used to reconstruct J/ψ mesons:

1. $J/\psi \to e^+ e^-$
2. $J/\psi \to \mu^+ \mu^-$

Only hadronic J/ψ candidates with an invariant mass between 2.8 and 3.5 GeV are considered.

B.1.3.8 D^{+*} Charged Spin-Excited D Meson from Decay-Channels

The following channels are used to reconstruct D^{+*} mesons:

1. $D^{+*} \to D^0 \pi^+$
2. $D^{+*} \to D^+ \pi^0$
3. $D^{+*} \to D^+ \gamma$

Only D^{+*} candidates with a released energy in the decay between 0 and 0.3 GeV are considered. Separate particle lists are created for D^{+*} meson reconstructed from hadronic, semileptonic and klong D mesons. If not stated otherwise D^{+*} always refers to a D^{+*}, which decayed in one of the hadronic decay modes listed above, throughout the text.

B.1.3.9 $D^*(2010)^0$ Neutral spin-excited D Meson from Decay-Channels

The following channels are used to reconstruct $D^*(2010)^0$ mesons:

1. $D^*(2010)^0 \rightarrow D^0\pi^0$
2. $D^*(2010)^0 \rightarrow D^0\gamma$

Only $D^*(2010)^0$ candidates with a released energy in the decay between 0 and 0.3 GeV are considered. Separate particle lists are created for $D^*(2010)^0$ meson reconstructed from hadronic, semileptonic and klong D mesons. If not stated otherwise $D^*(2010)^0$ always refers to a $D^*(2010)^0$, which decayed in one of the hadronic decay modes listed above, throughout the text.

B.1.3.10 D_S^+ Strange D Meson from Hadronic Decay-Channels

The following hadronic channels are used to reconstruct D_S^+ mesons:

1. $D_S^+ \rightarrow K^+K_S^0$
2. $D_S^+ \rightarrow K^+\pi^+\pi^-$
3. $D_S^+ \rightarrow K^+K^-\pi^+$
4. $D_S^+ \rightarrow K^+K^-\pi^+\pi^0$
5. $D_S^+ \rightarrow K^+K_S^0\pi^+\pi^-$
6. $D_S^+ \rightarrow K^-K_S^0\pi^+\pi^+$
7. $D_S^+ \rightarrow K^+K^-\pi^+\pi^+\pi^-$
8. $D_S^+ \rightarrow \pi^+\pi^+\pi^-$
9. $D_S^+ \rightarrow K_S^0\pi^+$
10. $D_S^+ \rightarrow K_S^0\pi^+\pi^0$

Only hadronic D_S^+ candidates with an invariant mass between 1.68 and 2.1 GeV are considered. If not stated otherwise D_S^+ always refers to a D_S^+, which decayed in one of the hadronic decay modes listed above, throughout the text.

B.1.3.11 D_S^+ Strange D Meson from Decay-Channels with K_L^0 Neutral Long-Lived kaons

Due to the different kinematic properties, D_S^+ mesons from decay-channels which include a K_L^0 are processed separately by the `FEI`. The following channels are used to reconstruct D_S^+ mesons:

1. $D_S^+ \rightarrow K^+K_L^0$
2. $D_S^+ \rightarrow K^+K_L^0\pi^+\pi^-$
3. $D_S^+ \rightarrow K^-K_L^0\pi^+\pi^+$
4. $D_S^+ \rightarrow K_L^0\pi^+$
5. $D_S^+ \rightarrow K_L^0\pi^+\pi^0$

There are no further cuts applied to D_S^+ candidates with K_L^0.

B.1.3.12 D_S^{+*} Strange Spin-Excited D Meson from Decay-Channels

The following channels are used to reconstruct D_S^{+*} mesons:

1. $D_S^{+*} \rightarrow D_S^+\gamma$ 2. $D_S^{+*} \rightarrow D_S^+\pi^0$

Only D_S^{+*} candidates with a released energy in the decay between 0 and 0.3 GeV are considered. Separate particle lists are created for D_S^{+*} meson reconstructed from hadronic D_S^+ and klong D_S^+ mesons. If not stated otherwise D_S^{+*} always refers to a D_S^{+*}, which decayed in one of the hadronic decay modes listed above, throughout the text.

B.1.4 B Mesons

All B mesons have the same input features for the multivariate classifiers: kinematic and vertex variables excluding quantities correlated to the beam-constrained mass; kinematic, tracking and vertex variables of all daughters evaluated in the rest frame of the B meson; the decay mode identifier and signal probability σ of all daughters; and the position of the candidate in the best candidate ranking of the pre-cut.

B.1.4.1 B^+ Charged B Meson from Hadronic Decay-Channels

The following hadronic channels are used to reconstruct B^+ mesons:

1. $B^+ \rightarrow \overline{D}^0\pi^+$
2. $B^+ \rightarrow \overline{D}^0\pi^+\pi^0$
3. $B^+ \rightarrow \overline{D}^0\pi^+\pi^0\pi^0$
4. $B^+ \rightarrow \overline{D}^0\pi^+\pi^+\pi^-$
5. $B^+ \rightarrow \overline{D}^0\pi^+\pi^+\pi^-\pi^0$
6. $B^+ \rightarrow \overline{D}^0D^+$
7. $B^+ \rightarrow \overline{D}^0D^+K_S^0$
8. $B^+ \rightarrow \overline{D}^{0*}D^+K_S^0$
9. $B^+ \rightarrow \overline{D}^0D^{+*}K_S^0$
10. $B^+ \rightarrow \overline{D}^{0*}D^{+*}K_S^0$
11. $B^+ \rightarrow \overline{D}^0D^0K^+$
12. $B^+ \rightarrow \overline{D}^{0*}D^0K^+$
13. $B^+ \rightarrow \overline{D}^0D^*(2010)^0K^+$
14. $B^+ \rightarrow \overline{D}^{0*}D^*(2010)^0K^+$
15. $B^+ \rightarrow D_S^+\overline{D}^0$
16. $B^+ \rightarrow \overline{D}^{0*}\pi^+$
17. $B^+ \rightarrow \overline{D}^{0*}\pi^+\pi^0$
18. $B^+ \rightarrow \overline{D}^{0*}\pi^+\pi^0\pi^0$
19. $B^+ \rightarrow \overline{D}^{0*}\pi^+\pi^+\pi^-$
20. $B^+ \rightarrow \overline{D}^{0*}\pi^+\pi^+\pi^-\pi^0$
21. $B^+ \rightarrow D_S^{+*}\overline{D}^0$
22. $B^+ \rightarrow D_S^+\overline{D}^{0*}$
23. $B^+ \rightarrow \overline{D}^0K^+$
24. $PBp \rightarrow D^-\pi^+\pi^+$
25v $B^+ \rightarrow D^-\pi^+\pi^+\pi^0$
26. $B^+ \rightarrow J/\psi K^+$
27. $B^+ \rightarrow J/\psi K^+\pi^+\pi^-$
28. $B^+ \rightarrow J/\psi K^+\pi^0$
29. $B^+ \rightarrow J/\psi K_S^0\pi^+$

Only hadronic B^+ candidates with an beam-constrained mass above 5.2 GeV and an absolute deviation of the energy from the beam energy below 0.5 GeV are considered.

B.1.4.2 B^+ Charged B Meson from Semileptonic Decay-Channels

Due to the different kinematic properties, B^+ mesons from semileptonic decay-channels are processed separately by the FEI. The following semileptonic channels are used to reconstruct B^+ mesons:

1. $B^+ \to \overline{D}^0 e^+\nu$
2. $B^+ \to \overline{D}^0 \mu^+\nu$
3. $B^+ \to \overline{D}^{0*} e^+\nu$
4. $B^+ \to \overline{D}^{0*} \mu^+\nu$
5. $B^+ \to D^-\pi^+ e^+\nu$
6. $B^+ \to D^-\pi^+ \mu^+\nu$
7. $B^+ \to D^{+*}\pi^+\mu^+\nu$
8. $B^+ \to D^{+*}\pi^+ e^+\nu$

In addition all hadronic channels listed in Sect. B.1.4.1 are used, where one of the hadronic D mesons is replaced by a semileptonic D meson. There are no further cuts applied to semileptonic B^+ candidates.

B.1.4.3 B^+ Charged B Meson from Decay-Channels with K_L^0 Neutral Long-Lived kaons

Due to the different kinematic properties, B^+ mesons from decay-channels which include a K_L^0 are processed separately by the FEI. All hadronic channels listed in Sect. B.1.4.1 are used, where either one of the hadronic D mesons is replaced by a klong D meson, or one of the K_S^0 is replaced by a K_L^0. There are no further cuts applied to B^+ candidates with K_L^0.

B.1.4.4 B^0 Neutral B Meson from Hadronic Decay-Channels

The following hadronic channels are used to reconstruct B^0 mesons:

1. $B^0 \to D^-\pi^+$
2. $B^0 \to D^-\pi^+\pi^0$
3. $B^0 \to D^-\pi^+\pi^0\pi^0$
4. $B^0 \to D^-\pi^+\pi^+\pi^-$
5. $B^0 \to D^-\pi^+\pi^+\pi^-\pi^0$
6. $B^0 \to \overline{D}^0\pi^+\pi^-$
7. $B^0 \to D^-D^0K^+$
8. $B^0 \to D^-D^*(2010)^0K^+$
9. $B^0 \to D^{+*}D^0K^+$
10. $B^0 \to D^{+*}D^*(2010)^0K^+$
11. $B^0 \to D^-D^+K_S^0$
12. $B^0 \to D^{+*}D^+K_S^0$
13. $B^0 \to D^-D^{+*}K_S^0$
14. $B^0 \to D^{+*}D^{+*}K_S^0$
15. $B^0 \to D_S^+D^-$
16. $B^0 \to D^{+*}\pi^+$
17. $B^0 \to D^{+*}\pi^+\pi^0$
18. $B^0 \to D^{+*}\pi^+\pi^0\pi^0$
19. $B^0 \to D^{+*}\pi^+\pi^+\pi^-$
20. $B^0 \to D^{+*}\pi^+\pi^+\pi^-\pi^0$
21. $B^0 \to D_S^{+*}D^-$
22. $B^0 \to D_S^+D^{+*}$
23. $B^0 \to D_S^{+*}D^{+*}$
24. $B^0 \to J/\psi K_S^0$
25. $B^0 \to J/\psi K^+\pi^-$
26. $B^0 \to J/\psi K_S^0\pi^+\pi^-$

Only hadronic B^0 candidates with an beam-constrained mass above 5.2 GeV and an absolute deviation of the energy from the beam energy below 0.5 GeV are considered.

The channel $B^0 \to \overline{D}^0\pi^0$ was used by the FR, but is not yet used in the FEI due to unexpected technical restrictions in the KFitter algorithm.

B.1.4.5 B^0 Neutral B Meson from Semileptonic Decay-Channels

Due to the different kinematic properties, B^0 mesons from semileptonic decay-channels are processed separately by the FEI. The following semileptonic channels are used to reconstruct B^0 mesons:

1. $B^0 \to D^- e^+ \nu$
2. $B^0 \to D^- \mu^+ \nu$
3. $B^0 \to D^{+*} \mu^+ \nu$
4. $B^0 \to D^{+*} e^+ \nu$
5. $B^0 \to \overline{D}^0 \pi^- e^+ \nu$
6. $B^0 \to \overline{D}^0 \pi^- \mu^+ \nu$
7. $B^0 \to \overline{D}^{0*} \pi^- e^+ \nu$
8. $B^0 \to \overline{D}^{0*} \pi^- \mu^+ \nu$

In addition all hadronic channels listed in Sect. B.1.4.4 are used, where one of the hadronic D mesons is replaced by a semileptonic D meson. There are no further cuts applied to semileptonic B^0 candidates.

B.1.4.6 B^0 Neutral B Meson from Decay-Channels with K_L^0 Neutral Long-Lived kaons

Due to the different kinematic properties, B^0 mesons from decay-channels which include a K_L^0 are processed separately by the FEI. All hadronic channels listed in Sect. B.1.4.4 are used, where either one of the hadronic D mesons is replaced by a klong D meson, or one of the K_S^0 is replaced by a K_L^0. There are no further cuts applied to B^0 candidates with K_L^0.

B.2 Automatic Reporting

B.2.1 Belle

Excerpt of the Full Event Interpretation Report (FEIR) for a generic training using 200 million Monte Carlo simulated $\Upsilon(4S)$ events (Tables B.2 and B.3).

B.2.2 Belle II

Excerpt of the FEIR for training using 180 million Monte Carlo simulated $\Upsilon(4S)$ events (Tables B.4 and B.5).

Table B.2 Per-particle efficiency before and after the applied pre- and post-cut

Particle	Pre-cut		Post-cut		
	User	Ranking	Absolute	Ranking	Unique
π^+	0.788	0.788	0.788	0.779	0.773
K^+	0.755	0.755	0.750	0.750	0.746
μ^+	0.892	0.881	0.865	0.853	0.852
e^+	0.775	0.769	0.757	0.757	0.754
γ	0.640	0.640	0.640	0.640	0.632
π^0	0.571	0.433	0.427	0.398	0.397
K_S^0	0.459	0.457	0.444	0.443	0.415
J/ψ	0.086	0.086	0.086	0.086	0.086
D^0	0.140	0.134	0.128	0.125	0.123
D^+	0.104	0.100	0.093	0.092	0.091
D_s^+	0.071	0.069	0.052	0.052	0.051
D^{0*}	0.053	0.051	0.048	0.047	0.047
D^{+*}	0.050	0.049	0.049	0.049	0.047
D_s^{+*}	0.035	0.034	0.031	0.031	0.030
B^0	0.005	0.005	0.005	0.005	0.004
B^+	0.008	0.008	0.008	0.007	0.007
$D^0{}_{SL}$	0.056	0.055	0.052	0.051	0.051
$D^+{}_{SL}$	0.059	0.058	0.052	0.051	0.050
$D^{0*}{}_{SL}$	0.023	0.020	0.015	0.015	0.015
$D^{+*}{}_{SL}$	0.022	0.022	0.020	0.020	0.020
$B^0{}_{SL}$	0.022	0.021	0.021	0.021	0.020
$B^+{}_{SL}$	0.020	0.019	0.018	0.018	0.018

Table B.3 Per-particle purity before and after the applied pre- and post-cut

Particle	Pre-cut		Post-cut		
	User	Ranking	Absolute	Ranking	Unique
π^+	0.639	0.639	0.696	0.713	0.707
K^+	0.137	0.137	0.394	0.394	0.392
μ^+	0.046	0.047	.086	0.094	0.094
e^+	0.050	0.052	0.193	0.195	0.194
γ	0.587	0.587	0.652	0.652	0.645
π^0	0.044	0.102	0.133	0.187	0.186
K_S^0	0.067	0.068	0.210	0.210	0.197
J/ψ	0.109	0.109	0.207	0.207	0.207
D^0	0.001	0.002	0.018	0.027	0.026
D^+	0.000	0.001	0.015	0.016	0.015
D_s^+	0.000	0.000	0.014	0.014	0.014
D^{0*}	0.000	0.001	0.011	0.012	0.011

(continued)

Table B.3 (continued)

Particle	Pre-cut		Post-cut		
	User	Ranking	Absolute	Ranking	Unique
D^{+*}	0.001	0.001	0.051	0.051	0.050
D_s^{+*}	0.000	0.000	0.010	0.010	0.009
B^0	0.000	0.000	0.000	0.001	0.001
B^+	0.000	0.000	0.000	0.001	0.001
$D^0{}_{SL}$	0.001	0.001	0.012	0.014	0.014
$D^+{}_{SL}$	0.001	0.001	0.007	0.007	0.007
$D^{0*}{}_{SL}$	0.000	0.000	0.005	0.005	0.005
$D^{+*}{}_{SL}$	0.000	0.000	0.012	0.012	0.012
$B^0{}_{SL}$	0.000	0.000	0.000	0.001	0.001
$B^+{}_{SL}$	0.000	0.000	0.000	0.001	0.001

Table B.4 Per-particle efficiency before and after the applied pre- and post-cut

Particle	Pre-cut		Post-cut		
	User	Ranking	Absolute	Ranking	Unique
π^+	0.757	0.757	0.756	0.747	0.737
K^+	0.793	0.793	0.790	0.790	0.782
μ^+	0.854	0.844	0.818	0.813	0.809
e^+	0.728	0.726	0.716	0.715	0.708
γ	0.579	0.579	0.579	0.579	0.577
π^0	0.349	0.342	0.341	0.326	0.325
K_S^0	0.425	0.425	0.269	0.269	0.263
J/ψ	0.091	0.091	0.091	0.091	0.090
D^0	0.145	0.139	0.134	0.129	0.127
D^+	0.092	0.088	0.082	0.082	0.080
D_s^+	0.063	0.060	0.054	0.053	0.052
D^{0*}	0.042	0.038	0.034	0.034	0.033
D^{+*}	0.047	0.047	0.046	0.046	0.045
D_s^{+*}	0.038	0.036	0.031	0.031	0.031
B^0	0.004	0.004	0.004	0.004	0.004
B^+	0.007	0.007	0.007	0.007	0.007
$D^0{}_{SL}$	0.064	0.063	0.060	0.059	0.058
$D^+{}_{SL}$	0.103	0.102	0.096	0.093	0.090
$D^{0*}{}_{SL}$	0.019	0.015	0.011	0.011	0.011
$D^{+*}{}_{SL}$	0.023	0.023	0.021	0.021	0.021
$B^0{}_{SL}$	0.023	0.022	0.022	0.021	0.020
$B^+{}_{SL}$	0.017	0.016	0.016	0.015	0.015

Table B.5 Per-particle purity before and after the applied pre- and post-cut

Particle	Pre-cut		Post-cut		
	User	Ranking	Absolute	Ranking	Unique
π^+	0.606	0.606	0.705	0.734	0.723
K^+	0.127	0.127	0.500	0.500	0.495
μ^+	0.042	0.046	0.106	0.112	0.111
e^+	0.050	0.055	0.213	0.216	0.214
γ	0.570	0.570	0.644	0.644	0.642
π^0	0.144	0.164	0.181	0.228	0.228
K_S^0	0.025	0.025	0.146	0.146	0.143
J/ψ	0.182	0.182	0.233	0.233	0.232
D^0	0.001	0.003	0.018	0.028	0.027
D^+	0.000	0.001	0.014	0.016	0.015
D_s^+	0.000	0.000	0.011	0.011	0.011
D^{0*}	0.000	0.001	0.007	0.008	0.007
D^{+*}	0.001	0.001	0.053	0.053	0.051
D_s^{+*}	0.000	0.001	0.007	0.007	0.007
B^0	0.000	0.001	0.001	0.001	0.001
B^+	0.000	0.000	0.000	0.001	0.001
$D^0{}_{SL}$	0.001	0.002	0.014	0.016	0.016
$D^+{}_{SL}$	0.002	0.002	0.009	0.012	0.012
$D^{0*}{}_{SL}$	0.000	0.000	0.005	0.005	0.005
$D^{+*}{}_{SL}$	0.000	0.000	0.015	0.015	0.015
$B^0{}_{SL}$	0.000	0.000	0.000	0.001	0.001
$B^+{}_{SL}$	0.000	0.000	0.000	0.001	0.001

Curriculum Vitae

Thomas Keck is an experimental high-energy physicist and machine-learning researcher. He obtained is Ph.D. in Physics at the Karlsruhe Institute for Technology in Germany. As a member of the Belle and Belle II collaboration he was responsible for the development and implementation of multivariate analysis methods in the Belle II Software Framework. In particular, his work was focused on hadronic and semileptonic tagging algorithms, and their application to rare B meson decays. His professional interests include any new technologies in the field of computer science in particular deep learning techniques and their application in science.

Publications

2018 B2BII: Data conversion from Belle to Belle II
Computing and Software for Big Science

2017 FastBDT: A speed-optimized multivariate classification algorithm for the Belle II experiment
Computing and Software for Big Science

2015 Optimierung der vollständigen Ereignis Interpretation von Belle II auf die Signalseitenselektion von Physik-Analysen
DPG Frühjahrstagung in Wuppertal

2014 An automated framework for hierarchical reconstruction of B mesons at the Belle II experiment
16th International workshop on Advanced Computing and Analysis Techniques in physics

T. Keck, *Machine Learning at the Belle II Experiment*, Springer Theses,
https://doi.org/10.1007/978-3-319-98249-6

Evaluation von multivariaten Klassifikationsmethoden für die vollständige Rekonstruktion bei Belle II
DPG Frühjahrstagung in Mainz

2013 Voxel2MCNP: A framework for modeling, simulation and evaluation of radiation transport scenarios for Monte Carlo codes
Physics in Medicine and Biology